T3-BQR-977

THE ORIGINS OF THE NUMBER CONCEPT

Charles J. Brainerd

PRAEGER PUBLISHERS
Praeger Special Studies

New York • London • Sydney • Toronto

Library of Congress Cataloging in Publication Data

Brainerd, Charles J
The origins of the number concept.

Bibliography: p.
Includes index.
1. Number concept. I. Title.
BF456.N7B7 512'.7'019 78-21223

ISBN 0-03-029306-5

PRAEGER PUBLISHERS
PRAEGER SPECIAL STUDIES
383 Madison Avenue, New York, N.Y. 10017, U.S.A.

Published in the United States of America in 1979
by Praeger Publishers,
A Division of Holt, Rinehart and Winston, CBS, Inc.

9 038 987654321

Printed in the United States of America

For my parents

God made the integers; all else is the work of man.

Leopold Kronecker

I have wished to understand the hearts of men. I have wished to know why the stars shine. And I have tried to apprehend the Pythagorean power by which numbers hold sway above the flux. A little of this, but not much, I have achieved.

Bertrand Russell

In the thought of number all the power of knowledge seems contained, all possibility of the logical determination of the sensuous. If there were no number, nothing could be understood, in things, either in themselves or in their relations to each other.

Ernst Cassirer

PREFACE

This book has two general aims. The first and more important objective, which occupies the entire volume, is to outline a unified approach to the origins of numerical concepts. The approach emphasizes the cultural evolution of numbers, logical theories of number, and what we know about how humans first come to grasp numerical ideas during the course of psychological development. Each of these topics has hitherto been considered an island unto itself. There are sound reasons for this view, based on the obvious differences in the methods employed to study each topic. These reasons notwithstanding, it has been my personal impression that one's understanding of each topic is deepened by considering it in relation to the other two. Unfortunately, the cultural history of numbers, logical theories of number, and the ontogenetic aspects of numerical reasoning are all vast fields of inquiry on which more than one book has been written. The literature on the third topic, in particular, has been rapidly proliferating in recent years. Therefore, it would be unreasonable to suggest that the present volume does complete justice to any of these topics. It does no more than set forth the boundaries of each domain and the principal lines of investigation.

My second objective, which occupies Chapters 6 through 11, is to propose a theory about ontogenetic changes in number concepts. The theory, which was originally suggested in some papers published in 1973, is examined in Chapter 6. Actually, the term "theory" is more than a little gratuitous. It is not so much a theory as a framework which requires constant tinkering as new data come in. Since the available data are as yet thin, the theory is necessarily primitive. Evidence relevant to the theory's empirical predictions is discussed in Chapters 7 through 10. Experiments conducted by other investigators are reviewed in Chapter 7. Since most existing research does not provide data relevant to the theory's predictions, only a fraction of the rather extensive literature on the development of number concepts is reviewed. Although I have endeavored to include all studies which provide tests of the theory's predictions, there have undoubtedly been oversights. Such omissions, though admittedly unfortunate, are scarcely avoidable. I can only apologize to those investigators whose work was not cited and assure them that the omission was unintentional. In Chapters 8 through 10, a series of experiments conducted in my own laboratories and at the University of Wisconsin's Research and Development Center for Cognitive Learning are reported. Broadly speaking, the findings tend to be consistent with theoretical predictions. Finally, in Chapter 11, the most obvious educational ramifications of the material presented up to that point are examined. If the theory proposed in Chapter 6 happens to be correct and if the main findings reported in Chapters 8 through 10 prove to be replicable, then the educational ramifications are, in view of current public school mathematics curricula, devastating.

Some caveats are in order regarding the discussion of logical theories of number in Part I. Since this book is written for a broad readership within the behavioral sciences and education, it seemed quite unreasonable to assume much familiarity with the logical foundations of arithmetic. Consequently, the material appearing in Part I is, to say the least, elementary. The style of presentation is both informal and inelegant—that is, I have not said anything in symbols and equations that could be said in words. Moreover, I have discussed only those matters which seemed indispensable to understanding the issues at stake in Part II. These features of Part I cause three problems which could easily be viewed as heinous by readers who are logicians or mathematicians by training. First, certain points which are the subject of debate in the technical literature are passed over. The definition of "class" in Chapter 6 is a case in point. In Principia, Whitehead and Russell maintained that classes could be identified with propositional functions in one unknown. Quine, on the other hand, maintains that this definition is too restrictive and that, for want of a formal definition, a class must be characterized as anything which makes a propositional function in one unknown true. I have simply adopted Quine's usage without stating my reasons. Second, some concepts which do not have the same meaning for mathematicians and logicians do have more or less the same meaning in Part I. An illustration is provided by the concepts of "series" and "progression." These notions have somewhat different meanings in abstract algebra and analysis, but I have used them interchangeably. Third, there is not a single "definition" in Part I that mathematicians and logicians would regard as precise—let alone elegant. The definitions of ordinal number (Chapter 3) and cardinal number (Chapter 4) are good examples.

While I acknowledge that Part I suffers from these shortcomings, I do not believe anything can be done about them. Since I am chiefly interested in communicating with behavioral scientists and educators, it seemed unwise to sacrifice comprehensibility for the sake of rigor. Moreover, if I had elected to write a treatment of logical theories of number that was both elementary enough to be understood by the readership to which this book is addressed and rigorous enough to satisfy those already acquainted with the subject, the result would have been a separate textbook on the logical foundations of arithmetic.

A profusion of such books is already available. Therefore, I would strongly suggest that readers who are already familiar with what modern logic has to say about numbers should not bother to read Part I. Instead, after Chapter 1, proceed immediately to Part II.

Although I should like to do so, it is impossible for me to acknowledge all the people who have shaped the views presented in this book. This is especially true of the material presented in Part II. I have discussed this material with so many people at scientific conferences, in telephone conversations, and in written correspondence that it is no longer possible for me to attach specific names to specific ideas. I would, however, like to mention those people who were kind enough to read the manuscript before it went to press: Edward H. Cornell, Rochel Gelman, Frank H. Hooper, Keith J. Laidler, Eugene S. Lechelt, Flimer S. C. Northrop, and Linda S. Siegel. Of these, I should like to single out Dr. Hooper, Dr. Laidler, and Dr. Siegel for special acknowledgement. Dr. Hooper prepared a detailed chapter-by-chapter critique of the original draft of the manuscript which proved to be of inestimable assistance when it came time to revise the manuscript. Dr. Hooper and his co-workers have also conducted some experiments designed to determine whether my own findings on the development of numerical ideas can be replicated. Dr. Laidler prepared a page-by-page list of typographical errors and grammatical violations which greatly facilitated correction of the typescript. Through his own writings, Dr. Laidler has made me aware of the pedagogical implications of the research reported in Part II. Without the stimulation provided by his writings, I seriously doubt that I should have written Chapter 11. Dr. Siegel and I, in the course of countless letters and conversations, have begun to formulate some tentative hypotheses which may explain the principal findings reported in Part II. We are currently in the process of conducting a series of follow-up experiments with normal and deaf children to determine whether these hypotheses are tenable. If all goes well, we shall report our findings in a sequel to the present volume.

I should also like to thank the following members of the cardinal-ordinal working group of the Number and Measurement Workshop: Morris Beers, James Hirstein, Martin Johnson, Curtis Spikes, Leslie Steffe, Robert Underhill, and Grayson Wheatley. This working group, which was originally formed in April of 1975, is affiliated with the University of Georgia's Center for the Study of Learning and Teaching Mathematics. One of the group's main goals is to explore the empirical predictions of the theory discussed in Part II and to apply the findings to public school mathematics curricula. This group's work has provided an invaluable stimulus to my own thinking about the problems of number development in children. The members' many trenchant criticisms of my views and their ingenious experiments have been extremely helpful. Importantly, they have caused me to rethink some of my ideas and, on various questions, have prevented me from sliding into the abyss of obscurantism.

Finally, I should like to express my thanks to the *Journal of General Psychology* for permission to reprint the data in Tables 8.1 through 8.5 and to the *Journal of Experimental Child Psychology* for permission to reprint the data in Table 9.2. I should also like to thank *Scientific American* for permission to reproduce Figure 6.5.

CONTENTS

LIST OF TABLES

LIST OF FIGURES

1

PROLOGUE

> Step by step the immemorial magic of numbers has kept pace with unmystical science down through the centuries. If the patient investigation of numbers has aided the development of science and furthered such enlightenment as science may give, it has also perpetuated beliefs that but few tolerant men would call enlightened.
>
> E. T. Bell

The lore of numbers is one of humanity's most ancient and redoubtable mental preoccupations. It is widely held that no other idea in the history of human thought even approaches number's combination of intellectual impact and practical ramifications. Although most of us view them as nothing more than useful scientific and mathematical tools, numbers—like prophecy, sex, and other mysterious things—have their darker side. Credible opinion from cultural anthropologists and mathematical historians has it that numbers originated as objects of superstitious dread or, what amounts to the same thing, religious veneration. There is considerable evidence that some pre historic cults worshipped numbers as a powerful source of magic. More recently, between the dawn of recorded history and the latter half of the nineteenth century, doctrines about the ultimate meaning of numbers have routinely involved religious suppositions of some sort. But more on this later. For now, it is pertinent to observe only that our modern belief—that numbers are singularly unmysterious entities which just happen to be of inestimable use in balancing the checkbook and figuring out the cost of living—is of rather recent vintage. This attitude would strike our not too distant ancestors as the rankest sort of heresy.

Although the precise origins of this predilection for number lore must remain shrouded in the dim mists of prehistory, we can say this much with virtual certainty: It was evident at the beginnings of recorded history along the Nile and Euphrates (circa 3500 B. C. and

2500 B. C. , respectively) and, therefore, it presumably preceded recorded history by a few milleniums or so.

By their nature, human beings are delvers into beginnings and plumbers of meanings. As a rule, we never rest content with simply putting useful ideas to work; we also feel compelled to root out their meaning. Further, the more useful an idea is the more we worry about its meaning. Unfortunately, the nature of human knowledge is such that the pursuit of meaning poses enormous obstacles. Somewhat paradoxically, it turns out that it is much easier for us to be certain about the truth of an idea or statement than it is for us to be certain about the meaning of same.

THE QUESTION OF NUMBER

Humanity has been asking itself, "What are numbers?", at least since the time of Pythagoras (sixth century B. C.) and probably since the rise of Babylonian mathematics around 2500 B. C. , or earlier. As it was just suggested, this question has proved rather difficult to answer. Not the least of the problems associated with this seemingly straight-forward question is the fact that its intent is somewhat unclear. It happens that historically, the question has been assigned two very different interpretations. Throughout most of recorded history, "What are numbers?", has been taken to mean, "Where did numbers come from?" We shall call this the metaphysical interpretation of our question. Given this interpretation, two different answers suggest themselves: (1) Numbers have an independent existence of their own—that is, numbers have always been there and humans merely "discovered" them in much the same way they discovered the planets of our solar system or the New World; (2) Numbers were invented to perform certain useful functions in much the same way that the wheel, the light bulb, and the radio were invented. During the sixth century B. C. , the father of mathematics, Pythagoras, and his followers argued that (1) was the correct answer. This doctrine eventually won overwhelming acceptance throughout most of the then-civilized world. It was retained as an article of faith by most Western philosophers and mathematicians until almost the beginning of the twentieth century. During the late nineteenth and early twentieth centuries, however, leading philosopher-mathematicians such as Gottlob Frege, Guiseppe Peano, Alfred North Whitehead, and, most of all, Bertrand Russell concluded that answer (2) probably was correct. We shall have more to say about each of these men in subsequent chapters. For now, it is important to note only that their arguments seemed more convincing than the old Pythagorean ones and, as a result, the majority opinion today is that numbers were invented rather than discovered.

Concurrent with the demise of the Pythagorean doctrine, the nature of numbers began to be interpreted in an entirely new way. The question now was taken to mean, "What more fundamental concept or concepts is the notion of number derived from?" We shall call this the scientific interpretation of our question. According to the scientific interpretation, we must discover the more basic ideas upon which the concept of number is founded if we are to understand the meaning of statements such as, "Five plus two equal seven." Like the metaphysical interpretation, there are two distinct ways of answering the scientific version of "What are numbers?" First, one can approach the problem the way a mathematician would, via logical analysis of the number concept. That is, one could perform a reductive analysis of numerical propositions in mathematics in an attempt to isolate the basic ideas which are implicit in these propositions. Once the analysis was complete, these more basic ideas could be used to construct a nominal definition of the number concept. But one can also approach the question the way a scientist would by doing empirical research on human numerical reasoning. The focus of this second approach is on the emergence of numerical ideas during the course of psychological development. Its aim is to identify the more basic skills that children must acquire before they can understand numbers and make numerical inferences. Unlike the two answers to the metaphysical interpretation of our question, the mathematical and empirical approaches to the scientific interpretation of "What are numbers?" are complementary. Logical analysis may give us hints about where to focus empirical research and empirical research may give us hints about where to concentrate logical analysis.

In this book, we shall be concerned with both the logical and empirical approaches to the scientific interpretation of "What are numbers?" The key things that logical analysis has revealed about numerical propositions are reviewed in Part I. Research dealing with the acquisition of numerical concepts during psychological development is reviewed in Part II. Before turning to these matters, however, it will be useful to examine some historical facts about humanity's long preoccupation with numbers. While many of these facts are interesting in their own right, the main reason for reviewing them is to lend perspective to the chapters which follow.

THE CULTURAL HISTORY OF NUMBERS

In this section, we consider some of the more notable events in the evolution of numerical ideas throughout history. In view of the fact that whole volumes have been devoted to this topic by others, these remarks clearly are not exhaustive. The aim is only to give the reader a thumbnail sketch which includes what mathematical historians deem to be the chief milestones in number's cultural genesis and will provide some appreciation of the long history of the problems to be taken up in Parts I and II. For convenience, the important events in the cultural evolution of number may be grouped into three periods; ancient, medieval, and recent. The first period spans at least the four milleniums immediately preceding the birth of Christ. The contributions to number lore during this period came, as far as we know, from the classical civilizations of Egypt, Babylon, and Greece. The second period spans the millenium and one-half immediately following the birth of Christ. During this period, the main contributions came from Christian and Islamic scholars. The third period runs from the fifteenth century until the present. It is during this period that mathematics as we know it emerged, and our current ideas about number were formulated.

Ancient Period

In its most basic psychological sense, number appears to be a quality of sensible objects. For example, the numerousness of a set of objects is just as much a perceptible quality of reality as color, flavor, odor, height, weight, and so on. Before number can begin its evolution as a concept, however, an apparently simple assumption must be made. Explicitly, it is necessary to assume that number is a property that is independent of sensible objects. In other words, number refers to the sensible world but it is somehow more than this; this is what will be called the abstract attitude toward numbers. The importance of the abstract attitude is very great because it permits us to use numbers and to invent computational systems such as arithmetic and algebra without having to ground what we do in direct experience or observation. Also, the attitude permits us to use moderately large numbers such as 1,000 and very large numbers such as 1,000,000 as easily as we use the first few integers. Although all of us have had direct experience with small integers up to 10 or 20 (by counting our fingers, the coins in our pocket, our offspring, and so forth), none of us can claim extensive experience with many moderately large or very large numbers. Who among us has ever bothered to count all the seats in a football stadium or the registered voters in a city? And yet, we are capable of asserting with complete confidence that a given stadium seats 101,000 and that a certain city has 2,728,483 registered voters on a given date. Obviously, our faith in the truth of such statements derives from something other than direct, experiential contact with each of the seats or each of the voters.

Although the abstract view of numbers has been commonplace in civilized societies for at least five and one-half milleniums, this was not always true nor is it universally true among all cultures today. It is still possible to find tribes of primitive peoples in the Amazon basin and in the southern Pacific in which this attitude is absent. Research conducted on the numerical capacities of such tribes (for example, Wertheimer 1938) underscores the importance of an abstract view of numbers. To begin with, numbers larger than 5 or 6 apparently are never used by such people. Beyond 5 or 6, everything is simply "many" or "numberless." Moreover, there is a complete absence of any conception that numbers larger than the very small ones which may be directly experienced even exist.

The first clear historical evidence of the appearance of the abstract numerical attitude comes to us from the Egyptians of 5,500 years ago. The evidence is in the form of a scepter carved for some ruler, perhaps a pharaoh. The symbols on the scepter report that, as a result of a long-forgotten campaign against unknown foes, the ruler's possessions increased by some 120,000 slaves, 400,000 oxen, and 1,422,000 goats. It appears highly improbable that the ruler's accountants actually had direct contact with each person, ox, and goat. As E. T. Bell (1946) has suggested, it is much more likely that they selected some reasonably large number (say, 1,000) and then proceeded to estimate the number of groups of this size. In any case, the important point for our purposes is that the use of such large numbers implies that a great insight, that number may be confidently treated as a property apart from the real world, had been achieved sometime before 3500 B.C.

Although the achievements of the Egyptians are not to be minimized, there is no evidence that they developed a comprehensive and precise science of numerical computation. The basic problem seems to have been that they never managed to move beyond the everyday uses of numbers toward something more closely resembling a mathematical approach to number. In their culture, numbers apparently never were thought of as anything more than a practical means to achieve practical ends. Hence, despite the fact that the Egyptians made extensive use of very large numbers and had rough-and-ready methods of computing them, such as the method mentioned above, they seem not to have progressed much beyond this level. In particular, Egyptian civilization made little progress in the development of the two rudimentary branches of mathematics normally associated with numerical computation—arithmetic and algebra. The computational needs of Egypt's simple agrarian society did not require a systematic exposition of arithmetic and algebra. Rough computations were more or less adequate for determining land allotments, building pyramids, computing the number of days between successive floods of the Nile, and so on. Hence, the cultural evolution of number eventually passed out of Egypt and into the hands of the technologically more sophisticated civilization of Babylon.

Babylonian civilization flourished along the Euphrates river in Asia Minor during the second and third milleniums B. C. There is evidence, in the form of clay tablets deciphered during the late 1920s and early 1930s, that Babylonian merchant-mathematicians took the second decisive step in the evolution of number; the development of systems of arithmetic and algebra. There are two major reasons why Babylon and not Egypt gave us our first systematic treatment of these branches of mathematics. First and perhaps most important, Babylon was a trading nation whose economic survival hinged on being able to give precise answers to numerical questions. Recall the Egyptian scepter and its inscribed numbers. Just how far away from the true numbers of captured slaves, oxen, and goats were these nicely rounded figures? Unless the ruler's accountants had more than their share of luck, each of the inscribed numbers is off by several thousand units. Although the probability of being so far off in their computations apparently did not bother the Egyptians, these are unconscionable errors by modern elementary school standards. They also were much too large for the Babylonians to tolerate. A Babylonian trader had to know precisely how many camels or coins or tents or bushels of grain one slave would bring. An error of even one camel or coin or tent or bushel could be the difference between prosperity and ruin. The second reason why Babylon developed arithmetic and algebra while Egypt did not is more subtle. The Babylonians seem to have recognized an important fact about practical computation that escaped the Egyptians; before practical computation can attain the high degree of accuracy that nonagrarian civilization requires, rough-and-ready methods must be discarded, and calculation must be pursued as an abstract or "pure" science.

The Babylonians began working out a comprehensive exposition of arithmetic well before 2500 B. C. By this date, the merchant-mathematicians of the Euphrates valley had formulated an arithmetic of sufficient power to handle most trading problems concerned with money, weights and measures, and bartering. The Babylonian system of arithmetic—which unlike our own employed a base-60 number system—was fully articulated as an abstract computational science by about 2000 B. C. and was so powerful and comprehensive that more than four milleniums were to pass before European mathematicians were able to formulate a clearly superior system. Between 2000 B. C. and 1200 B. C. , Babylonian mathematicians concentrated on the development of algebra. Although the algebra developed during these eight centuries was far from complete by modern standards, it clearly surpassed all other systems of algebra formulated between 1200 B. C. and the European Renaissance. To illustrate their algebraic accomplishments, Babylonian mathematicians managed to discover most of the rules for solving quadratic equations. Given sufficient time, the Babylonian algebraist of 1200 B. C. could solve most of the story problems found in today's high school algebra textbooks. For the next 27 centuries, no other group of mathematicians would be able to make a similar claim.

Despite their monumental achievements in arithmetic and algebra, Babylonian mathematicians were not entirely scientific in their approach. On occasion, they gave in to superstition and obscurantism. In addition to formulating the objective mathematical properties of numbers, the Babylonians attributed magical powers to at least some numbers. To give but one example, there is evidence that they regarded numbers which may be evenly divided by many other numbers as magical. This darker side of Babylonian number lore foreshadows the work of the Greeks and early Christians. Both of these latter groups were prone to mix mathematics with liberal doses of mythology.

The many important differences between Egyptian and Babylonian number lore notwithstanding, the Egyptians and Babylonians seem to have agreed on one fundamental point. The number work of both groups was exclusively computational. Both groups were primarily concerned with putting numbers to work to solve problems. As we have seen, the Babylonians were much better at it than the Egyptians were. But this is a difference of amount rather than kind. Because their number work was problem-oriented, neither the Egyptians nor the Babylonians thought to construct a philosophy of number. In particular, there appears to be no evidence that the philosophical question "What are numbers?", aroused much interest. To the no-nonsense Egyptians, this question probably would have seemed quite meaningless. To the Babylonians, with their more abstract approach to number, the question may have had some meaning; however, if they posed the question, the more pressing problem of creating comprehensive computational sciences left the Babylonians little time to ponder it. Hence, although the Babylonians were the ancient world's masters of calculation, their philosophy of number was nonexistent. They discovered virtually all the common uses of numbers, but as to the meaning of these extremely useful entities they apparently said nothing.

It fell to the Greeks of the sixth to fourth centuries B. C. to construct the first successful philosophy of number. The Greek approach to numbers (and to mathematics in general) was the polar opposite of the Egyptian and Babylonian approaches. The Greeks were not overly concerned with either numerical computation or most other forms of working mathematics. Geometry was the only form of working mathematics in which they excelled. The Greeks founded their arithmetic and algebra on geometry, instead of the reverse as is common today. Although Greek arithmetic and algebra were computationally superior to the rough-and-ready methods of the Egyptians, they fell far short of the Babylonian standards of six centuries earlier.

What the Greeks excelled in was not so much mathematics as philosophy of mathematics. In fact, the Greeks are generally regarded as the originators of the philosophy of mathematics. Following Bertrand Russell (1919), we draw the boundary between mathematics and philosophy of mathematics roughly as follows: When doing mathematics, one engages in a process whereby the scope of application of mathematical notions is broadened. That is, one increases the variety of ways in which mathematical notions can be used. Doing mathematics, therefore, consists mainly in putting mathematical constructs to work in novel ways, as in proving new theorems. When doing philosophy of mathematics, on the other hand, one engages in the reverse process. Rather than expanding the scope of a mathematical notion, one narrows it. Explicitly, one attempts to reduce a complex mathematical construct which is known to have rich field of application to a comparatively simple definition. In philosophy of mathematics we ask, "What does this concept mean?" Greek mathematicians found questions of the latter sort more stimulating. In answer to one such question, "What are numbers?", they evolved a doctrine that stood virtually unchallenged for 25 centuries.

Here we consider only the two most important contributors to Greek philosophy of number: Pythagoras and Plato. Pythagoras was by far the more important of the two. He was born in or about 569 B. C. on the island of Samos and died in or about 500 B. C. in Italy. Pythagoras spent roughly the first 20 years of his life on Samos. During the next 20 or so years, he travelled widely and spent a considerable amount of time in the company of Egyptian and Babylonian philosophers. From the Babylonians, Pythagoras received his first exposure to the doctrine that numbers have magical properties. During his travels, Pythagoras also studied some of the exotic tenets of Eastern religions. When Pythagoras eventually returned to his native Samos, he began to preach doctrines such as reincarnation and transmigration of souls in a local open-air theater. His teachings were not well received by local guardians of the faith. Centuries before, Homer had created a mythology which had become an institutionalized religion by Pythagoras' time. Doctrines such as reincarnation and transmigration of souls were quite alien to Homerian mythology. Ultimately, Pythagoras' teachings were the subject of an inquisition of sorts. The verdict was inevitable. Fortunately for posterity, the Greeks were not as fond of torturing their heretics and burning them at the stake as the Christian fathers of later centuries. Under the admirably enlightened maxim that heresy at a very great distance is the same thing as no heresy at all, heretics usually were banished if they refused to repent. So it was with Pythagoras.

Pythagoras set sail with his aged mother for the Greek colony of Croton on the Italian peninsula sometime between the ages of 40 and 50. His fame as a teacher preceded him and he received a warm welcome from the provincial Crotonians. Croton was ruled by the most famous Olympic champion of the ancient world, Milo, of whom it is said that he once paraded around the stadium at Olympus with a bull on his shoulders, then killed the bull with his bare hands, and ate the poor beast raw. Milo became Pythgoras' patron and Pythagoras dwelt in Milo's home. It was on Milo's grounds that Pythagoras founded the ancient world's most famous academy for philosopher-mathematicians, the Pythagorean Brotherhood. The brotherhood flourished for about a quarter of a century under Milo's patronage. Its work was equal parts mathematics and superstition. The mathematics was primarily geometry (for example, the famous Pythagorean theorem) and the superstition was primarily numerical. Members of the brotherhood also meddled in politics. They were supporters of oligarchic forms of government and were horrified by all forms of democracy. This proved to be their undoing. Around 500 B. C., a political agitator and ex-member of the brotherhood, Cylon, led a mob of Crotonians in an assault on Milo's grounds. They set fire to Milo's house, and tradition has it that Pythagoras and most of his followers were trapped inside.

Sometime during his years on Croton, Pythagoras arrived at a startling answer to the question, "What are numbers?" Numbers, he decided, are quite literally all that there is. Number is the only true reality and numbers are the language of the universe. To understand how and why Pythagoras could arrive at such a sweeping generalization, it is necessary to recall an important fact about philosophy as the Greeks practiced it. Greek philosophy aimed at nothing less than comprehension of the entire universe and everything in it. The Greeks believed that the path to such comprehension lay in massive generalizations about the underlying substratum of the universe. In their view, the universe and everything in it could be reduced to one or a few such generalizations. In addition to Pythagoras' generalization about number, other popular Greek generalizations of this era included "all is water" and "all is fire." Hence, despite the fact that "all is number" seems more than a little naive today, it is an excellent example of the Greek approach to decoding nature's secrets. For our purposes, the crucial point to note about "all is number" is that this dogma entails that numbers exist—in the same sense that trees and rocks exist—and are not mere products of human imagination.

Pythagoras' working concept of number was much narrower than our own. Today, high school and college students understand by "number" such things as algebraic numbers, complex numbers, hypercomplex numbers, and transcendental numbers. None of these more exotic types of number were known during Pythagoras' time. In fact, Pythagoras acknowledged only two types of numbers: (1) the unending sequence of positive integers (1, 2, 3, . . .) that are called the "natural numbers" today and (2) the unending sequence of fractions (a_1/b, a_2/b_2, a_3/b_3, . . .) in which all numerators and denominators are natural numbers (called "rational numbers" today). When Pythagoras said, "All is number," he had only the natural numbers 1, 2, 3, . . . in mind. Historically, this statement produced two important offshoots—one a conjecture about the foundations of mathematics and the other a conjecture about the foundations of the physical universe. The first (and historically the more important) of these conjectures proposed that numbers are the final and absolutely irreducible core from which all mathematics is derived. In the Pythagorean view, mathematics could not go below the cornerstone of number. Thus, the scientific meaning of "what are numbers?"—where we take this phrase to mean a definition of number in terms of more basic notions—was a metaphysical impossibility for Pythagoras. The second notable offshoot of "all is number" proposed that natural numbers are the building blocks of the material universe. That is, matter itself is somehow built up from 1, 2, 3, ...

Even the rudimentary state of Greek physics must have made this second conjecture somewhat improbable. In fact, it probably was abandoned while Pythagoras still was alive. The principal reason for its abandonment was the Pythagorean Brotherhood's discovery that certain physical quantities cannot be exactly represented either by a natural number or by any ratio of natural numbers. Their doctrine that matter is built up from 1, 2, 3, . . . necessitated that all physical quantities should be capable of exact representation as either a natural or rational number. In contrast, the Pythagorean conjecture that numbers form the ultimate foundation of mathematics turned out to be far more resilient. It remained an unquestioned tenet of mathematical faith for roughly two and one-half milleniums after Pythagoras' death. As a result, it was universally believed until very recently that numbers are the only proper business of mathematics. The focus of working mathematics did not drift decisively away from number until the third decade of the nineteenth century. Throughout the long history of mathematics, no other doctrine can claim even a remotely comparable period of influence.

Plato will be the final figure in the litany of number lore in the ancient world. Plato was born in either 428 or 427 B. C on the island of Aegina and died in Athens in 347 B. C. During his early years, Plato concentrated on athletics (he was an Olympic games participant) and writing poetry. He met the philosopher Socrates in or about 407 B. C. Following this fortuitous meeting, Plato gave up athletics and poetry in favor of becoming Socrates' pupil. He remained his pupil until Socrates' execution in Athens in 399 B. C. Between 399 B. C. and 387

B C., Plato traveled extensively. During this time, he studied the philosophy of Pythagoras and the geometry of Euclid. Plato eventually returned to Athens and founded a famous academy. It was devoted to philosophy and politics, and it was frequented by most of the leading thinkers of the day, including Aristotle. Unlike Pythagoras' brotherhood, Plato's continued to function long after his death, surviving until the sixth century A. D.

While Pythagoras was equal parts mathematician and philosopher, Plato was all philosopher. He did no original mathematics of his own and relied on the authority of Euclid and Pythagoras in such matters. In his philosophy, Plato was a Pythagorean. This may surprise some readers. Thanks to his prolific writings, we tend to think of most of the important ideas in his works as being either original or deriving from Socrates. This is incorrect. There is one fundamental premise that underlies all of Plato's philosophy: Ideas, whether mathematical or physical or social, have an existence of their own that is quite independent of the human mind. The mind does not "have" ideas; it "observes" them. We already have seen that, at least in so far as numbers are concerned, this "Platonic" doctrine was the cornerstone of Pythagoras' credo. In any case, the point to be emphasized is that Plato reached more or less the same conclusion about the origins of numbers as Pythagoras—that is, he concluded that they are "real."

Plato's somewhat elliptical line of reasoning ran as follows: The quality of "number," unlike qualities such as odor, color, and taste, is very abstract. While we experience and gain knowledge about odors, colors, and tastes in a direct manner via our sense impressions, we do not seem to gain knowledge about numbers in this way. And yet, it is obvious that we know a great deal about numbers—perhaps even more than we know about qualities experienced via sensation. How is such knowledge obtained? The mind must obtain it without benefit of sensation by simply contemplating numbers. But if the mind contemplates things that cannot be sensed, then these things must exist independent of the mind and all other material things. So it is with number. Plato's belief about the independent existence of numbers in particular and mathematical concepts in general are contained in three of his most famous dialogues: The Meno, The Phaedo, and The Republic.

Medieval Period

There is little of significance to be said about number lore during the Middle Ages. It is symptomatic of this era that scientific and mathematical discoveries, as we currently understand them, were exceedingly rare. It is well known that this period of unparalleled dogma and devotion to armchair speculation was hardly conducive to such enlightment as science and mathematics may give. Consequently, the contributions of medieval number scholars are distinguished more by their curiosity value and obscurity than by their value as positive contributions to knowledge. For our purposes, the most important point about the Middle Ages is that the mystical pseudoscience of numerology held absolute sway over the number lore of this period. The work of most medieval number scholars (one could not reasonably call them mathematicians) began and ended with numerology.

Many readers, no doubt, are familiar with numerology as a form of party and carnival entertainment, similar in principle to palmistry, card reading, astrology, and crystal-ball gazing. For the sake of completeness, let us consider a brief illustration of modern numerology. A few years ago, an issue of the student newspaper at the University of Alberta (The Gateway, March 26, 1974) contained a numerological column written by Sybil Leek. According to Ms. Leek, who practices other forms of magic when not writing columns on numerology, numerical analysis of one's name can reveal hidden truths about one's personality and what the future holds. To discover these truths, we are instructed to proceed as follows. First, convert the letters in one's name to numbers via the following table:

1	2	3	4	5	6	7	8	9
A	B	C	D	E	F	G	H	I
J	K	L	M	N	O	P	Q	R
S	T	U	V	W	X	Y	Z	

Next, add these values and reduce the resulting number to the lowest possible number between 1 and 9 by either adding or subtracting the component digits. Finally, look up the result in another table provided by Ms. Leek to discover the truth about oneself. For example, consider the first great numerologist, Pythagoras. Converting the 10 letters into their respective numbers and adding them, we have 49. Reducing to the lowest possible number between 1 and 9 by subtracting 4 from 9, we have 5. Looking up 5 in Ms. Leek's table, we find that Pythagoras was a free spirit who loved change. History records that this is a somewhat improbable diagnosis.

Compared to the parlor-game example just considered, medieval numerology was very serious business. To understand medieval numerology, we shall require a more precise definition of the term "numerology" itself. By "numerology," we shall understand the thesis that the truth or falsity of any proposition whatsoever (statements about chemical compounds, statements about morality, statements about theology, and so on) can be demonstrated by numerical analysis. To prove which form of government is best, how RNA molecules translate the information in DNA molecules into biochemical results, and so on, the faithful numerologist has only to assign numbers to statements and perform arithmetic operations on the numbers. Numerology leads unfailingly to the more general doctrine that we can obtain valid information about the world without the onerous necessity of having to resort to the scientific method. While the scientist gradually builds up empirical data by patient experimentation, the numerologist solves nature's problems after lunch with paper and pencil. Of course, these claims seem wildly improbable from the lofty perspective of the second half of the twentieth century. During the Middle Ages, however, numerology was almost universally accepted as the method for obtaining valid knowledge about anything.

As might be expected from earlier remarks, both the Pythagorean Brotherhood and Plato's academy espoused forms of numerology. Important ideas such as "beauty," "god," "truth," "evil," "justice," "man," and "marriage" all had their special numbers. In fact, to be perfectly correct, important ideas were their special numbers in Pythagorean and Platonic philosophy. Both Pythagoras and Plato thought that the truths of the universe could be unraveled by doing arithmetic with such special numbers. But it was neither the Pythagorean Brotherhood nor Plato's academy that brought numerology to its full flower. This dubious honor fell first to Roman scholars and, subsequently, to Christian scholars.

The Roman Nigidius Figulus, who lived during the first century A. D. , is the acknowledged father of medieval numerology. Nigidius founded a school of philosophy called neo-Pythagoreanism; however, the only resemblance between Nigidius' teachings and the work of Pythagoras was numerology. Nigidius and his followers did not engage in serious mathematics at all. The most important function served by Nigidius' school was to spread Pythagorean and Platonic numerology beyond the recondite confines of Greek academies to the other civilized countries bordering the Mediterranean. Nigidius' only substantive contribution to numerology lay in extending it to the realm of superstition. It should be emphasized that the numerology of Pythagoras was concerned primarily with questions that would be viewed today as more or less legitimate scientific problems. Among other things, Pythagorean numerology dealt with questions about the physical structure of the universe that physicists and astronomers still continue to investigate—albeit by radically different methods. Nigidius' school, and a later Roman philosophical school known as neo-Platonism, found such questions rather dry and uninteresting. Theology was where their interests lay and they perceived great possibilities in numerology. Explicitly, they decided that numerology could be used to establish the truth of controversial theological propositions of the day.

Nigidius' school flourished from the first century B. C. until roughly the second century A. D. Then, until the fifth century A. D., control of medieval numerology passed to Alexandria. The Alexandrian numerologists of this era extended Nigidius' theologically oriented numerology to its logical extreme by inventing "gematria." In gematria, numerology is applied to the Old Testament scriptures. Verses in the Old Testament having certain numbers of letters, words, punctuation marks, and so forth are interpreted as containing hidden meanings of magical significance. Verses containing either direct or thinly veiled references to certain especially magical numbers (for example, 1, 3, 7, and 10) were also plumbed for hidden meaning. Readers familiar with the Bible will recall that there are many verses of this sort. In the Book of Revelation for example, there is the infamous "mark of the beast," the number 666. By adding or multiplying such numbers, hidden meanings were ostensibly revealed. Historically speaking, the invention of gematria by the Alexandrians is very important because a variant of gematria became the dominant form of numerology during the final two-thirds of the Middle Ages.

St. Augustine (353-430) appears to have been the first major convert to numerology among the early Christian fathers. The Christian church's advocacy of numerology during the remainder of the medieval period is due in large measure to St. Augustine's original efforts on its behalf. Augustine believed that gematria, applied to the New Testament rather than the old, was the one method by which Christian dogma could be proved to the satisfaction of all. This was an extremely important point in numerology's favor during the fourth and fifth centuries. At that time, the Christian church did not enjoy the complete dominion over the spiritual life of the West that it did a few centuries later. It was one of many competing faiths. Naturally, church officials were open to suggestions that would strengthen the church's influence. Augustine is reputed to have performed a complete gematria-like analysis of the New Testament, following which he concluded that God is a numerologist and numbers are a great common language which God has bestowed upon humanity so that it might fathom His designs. Pursuant to these momentous conclusions, Augustine bent his efforts to convincing church prelates that numerological methods should receive official sanction. Thanks to these efforts, serious Christian scholars soon were perpetrating a rich new assortment of numerological hoaxes on the New Testament, to which the mysterious title "transcendental arithmetic" ultimately was given.

Numerological analyses of the New Testament continued to prosper from the fifth century A. D. until the close of the Middle Ages. Perhaps the most popular object of these analyses was to compute the exact year in which the judgment day would occur. Christ had said that he would return swiftly but the centuries continued to pass; it would be nice if the faithful knew just how much longer they would have to wait. Other popular objects of Christian numerology were computing the exact year of creation and deriving numerological support from holy writ for Ptolemy's geocentric model of the universe.

Among the most famous post-Augustine numerologists were Proclus (411-485), St. Isadore (570-636), St. Thomas Aquinas (1226-1274), and the poet Dante (1265-1321). Proclus claimed to have invented numerological formulas so powerful that he could bend all physical and magical forces to his will. The wind, the tides, the sun, and spirits hastened to do Proclus' bidding. Unfortunately, Proclus did not have the forethought to base his numerological formulas on the New Testament and he was the subject of unrelenting persecution by the church. St. Isadore, a Spaniard, was a Christian numerologist in the Augustinian mold. He is reputed to have prepared a dictionary containing all the numerical references in both the Old and New Testaments. Aquinas expanded the Christian numerology of preceding centuries to include numerological ideas he discovered in Aristotle's works. Dante put numerology to poetic uses. He was fond of disguising potent numerological formulas in his verses. Dante's masterpiece, _The Divine Comedy_, is fraught with veiled numerical symbolism. For example, references to the magical number nine (so considered because it is the square of the Christian holy trinity) occur repeatedly.

During and shortly after the Protestant Reformation, a new subject of gematrical analysis became popular among Christian numerologists: proving that the pope (whoever he happened to be) was the Antichrist. Probably the most famous of these demonstrations appeared in 1593 in a book published by the Scot, John Napier. The book carried the incredible title _A plaine discovery of the whole revelation of Saint John: set downe in two treatises: The one searching and proving the true interpretation thereof: The other applying the same paraphrastically and historically to the text_. Napier usually is remembered as the inventor of logarithms, which frequently are viewed as the greatest labor-saving device ever invented. Most of us would not know how to find the decimal root of a number or raise a number to a decimal power without first converting to logarithms. In the book in question, however, Napier derived the date of the world's ending from a numerological proof that the pope is the Antichrist which he borrowed from Michael Stifel, an earlier German mathematician. Stifel managed to show that numerological analysis of the name of the sitting pope, Leo X, produced the number 666, the previously mentioned mark of the beast. In Latin, Stifel observed, Leo X is Leo Decimvs. The letters L, D, C, I, M, and V in this name also happen to be Roman numerals. Stifel proceeded to rearrange them to yield DCLXVI, the Roman equivalent of 666. To obtain this result, it was necessary to substitute X for M, which Stifel justified on the ground that X was

Leo's numerical designation. In his book, Napier argued that it was the pope of 1593, not Leo X, who was the Antichrist. Further numerological analysis led him to conclude that the world would end between 1688 and 1700. Unless we are deceived, Napier was in error.

Today, it is difficult to take seriously the flights of numerological fancy which characterized the Middle Ages. However, we would do well to remember that gematria, transcendental arithmetic, St. Isadore's dictionary, and the rest, no matter how far-fetched they may seem, were sober attempts by the leading scholars of the period to come to grips with what they believed to be fundamental problems. The mere fact that these efforts did not succeed or that they were predicated on a wildly improbable methodology does not make them any the less serious.

Before proceeding to the recent period, it should be noted in passing that the debasement of number lore into numerology was not quite total during the Middle Ages. Between the seventh and twelfth centuries, Moslem mathematicians pursued arithmetic and algebra in the manner of their Babylonian and Greek forebears. During these centuries, Bagdad was the capital of number science. Moslem mathematicians added the zero, which the Babylonians almost certainly had, to the natural numbers and rational numbers of the Greeks. They also gave us our modern number symbols. In view of what was going on in the rest of the civilized world, the accomplishments of the mathematicians of Bagdad often are overrated. Beyond the addition of zero to the Greek number systems and their new number symbols, Moslem mathematicians made few original contributions. They sought primarily to imitate and preserve the work of the Greeks, rather than to strike out in new directions of their own. Inevitably, their arithmetic and algebra were inferior to that of the Greeks and the Babylonians. In much of their work, the Moslems were hindered by a problem that plagued Greek arithmetic and algebra: the narrowness of their number concept. The Greeks, it will be recalled, accepted only the natural numbers and fractions derived from them. For complete systems of arithmetic and algebra, however, other forms of number must be admitted—in particular, negative numbers, irrational numbers, and complex numbers. The Greeks regarded these other forms of number as fictive and so did the Moslems.

Recent Period

The beginnings of the recent period in number lore can be traced to the advent of the European Renaissance. For convenience, this period may be subdivided into three phases. The first phase includes the fifteenth and sixteenth centuries plus the first three decades of the seventeenth century. The thesis that manipulating numbers is the appropriate method for obtaining valid knowledge was successfully challenged and discredited during this phase; the year 1633, when Galileo was tried and condemned for heresy, marks the nominal end of the first phase. The boundaries for the second phase are the fourth decade of the seventeenth century and the third quarter of the nineteenth century. During this time, the number concept of the Greeks was broadened to include many new forms of number which are obtained by performing arithmetic and algebraic operations on natural and rational numbers. Modern abstract arithmetic, abstract algebra, and analysis became possible because of these successive extensions of the number concept. The third phase began during the fourth quarter of the nineteenth century and it has been characterized by attempts to consolidate rather than expand the number concept. Explicitly, the third phase has witnessed attempts to construct a logically satisfactory definition of the system 1, 2, 3, . . . of natural numbers.

First Phase

The discrediting of numerology was largely a result of the dramatic events surrounding the lives and work of two men: Giordano Bruno (1548-1600) and Galileo (1564-1642). Bruno and Galileo both challenged important Christian dogmas whose truth supposedly had been established many times over by medieval numerology. In so doing, they undermined their contemporaries' faith in numerology itself.

Born in the very center of Christian numerology, Italy, Bruno was himself a devotee of numerology—but not Christian numerology. Bruno's numerology was cast more in the Pythagorean and Platonic molds. In the eyes of the church, therefore, he was pagan. Being of humble peasant stock, Bruno could not bring himself to countenance the wide variety of violations of common sense that Christian numerology had "proved," and he inveighed against many of them. Bruno's opposition to one doctrine in particular, the Ptolemaic model of the universe, is noteworthy. During the Middle Ages, Ptolemy's theory that the Earth is the center of the universe had been established to everyone's satisfaction via, among other things, numerological analysis of holy writ. However, Bruno taught that Copernicus' posthumously published heliocentric model of the universe was, in fact, correct. Bruno also launched some memorable attacks on the numerological eccentricities of Dante's *Divine Comedy*. The pope's holy inquisition sought to bring Bruno to trial on heresy charges for his teachings. Fortunately, Bruno foresaw this possibility and removed himself to Protestant Geneva, where, he took up his teaching again. His opinions eventually fell out of favor in Geneva, and he was forced to flee to

France and, subsequently, to England. In 1593, the same year in which Napier's book was published, Bruno made the mistake of returning to Italy for the purpose of resuming his teaching in Venice. The Venetians betrayed him into the hands of the holy inquisitors who imprisoned him in the hope that he would repent and confess his heresy. Bruno proved intractable, however. After seven years, the inquisitors' patience finally was exhausted; they burned Bruno at the stake in February of 1600.

The story of Galileo is no doubt better known than Bruno's. Unlike Bruno, Galileo was not a confirmed numerologist. Galileo believed that new and valid knowledge is obtained principally through empirical means—that is, by patient experimentation and careful observation. According to Galileo, the correct view of numbers and numerical formulas is not that they provide new knowledge but, rather, that they are post facto devices for representing new knowledge won through experimentation. Galileo's classic experiment involving the tower at Pisa illustrates this view. Since Plato's time, it had been obvious to all rational minds that if two objects of unequal weight are dropped from the same height, the heavier one will fall at a faster rate. Any required amount of numerological proof of this obvious fact could be easily provided. What possible need was there to resort to vulgar experimentation to confirm such a simple numerological truth? However, Galileo performed the experiment and, unbelievably, the two objects fell at the same rate. With this single rudimentary experiment, Galileo invalidated a proposition that numerology had accepted unquestioningly for two milleniums. More important than the disproof of this specific proposition was the fact that the experimental method had proved superior to numerological demonstration. Numerology was never to recover from this devasting blow.

Like Bruno, Galileo disbelieved the geocentric model of the universe and advocated Copernicus' alternative theory. More than any of his other teachings, this particular belief was to be Galileo's undoing in the eyes of the church. Papal representatives warned Galileo about the heretical implications of his teachings many times during the course of his career. Somehow, by dint of luck or wit, he always managed to avoid being brought to trial, even though he continued to live in Italy. Finally, in 1632, he published a philosophical disquisition which purported to be a neutral airing of the respective merits of the heliocentric and geocentric theories of the universe. However, it was obvious from the text that the overwhelming weight of evidence favored the heliocentric model. The inquisitors at last were moved to take action. They tried Galileo and found him guilty of heresy in 1633. Because Galileo was regarded as the leading scholar of his day, detailed records of the trial were kept which are available today in the Vatican library. Faced with the same choice as Bruno 33 years before, Galileo chose public confession of his heresy in preference to death at the stake. Following his confession and a declaration of repentance, he was placed under house arrest for the remaining nine years of his life.

From the perspective of the history of number lore, the importance of Bruno and Galileo does not lie in any minor battles that either may have won against orthodox Christian numerology. In the end, they both lost the decisive battle. However, the extreme measures, murder and persecution, that the church was forced to resort to in its attempts to silence Bruno and Galileo served to dramatize the absurdity of numerology in the minds of the educated and, ultimately, to foster deviancy rather than eliminate it. It probably is fair to say that the burning of Bruno and the persecution of Galileo did more to free number lore from the yoke of numerology than anything that either of them said or could have hoped to say.

Second Phase

The number concept of the ancient and medieval periods was quite restricted by modern standards in that only the natural numbers and the fractions which may be constructed from them were accepted as legitimate numbers. During the second phase of the recent period, the number concept was broadened to include new types of numbers which result from the application of arithmetic and algebraic operations to natural and rational numbers. We shall consider the development of two of these new types of number whose cases are fairly typical: negative numbers and complex numbers. Where a and b are natural numbers or rational numbers and $a < b$, negative numbers result from expressions of the form $(a - b)$. Where a and b are negative numbers and i is any even natural number, complex numbers (also called "imaginary numbers") result from expressions of the form $i/\sqrt{a}$ and $i/\sqrt{b}$. Negative and complex numbers are important because complete systems of arithmetic and algebra, respectively, are impossible without them. For arithmetic to be complete, we must admit all the numerical consequences of arithmetic operations (in this case, subtraction). For algebra to be complete, we must admit all the numerical consequences of algebraic operations (in this case, taking even integral roots).

The development of negative numbers went through three stages. First, they were discovered as a natural consequence of applying the subtraction operation to accepted types of numbers, but they were denied the status of legitimate numbers and remained largely unused. Second, while their status as legitimate numbers continued to be denied, their use in arithmetic computation became extensive. Finally, when arithmetic had become so thoroughly permeated with negative numbers that it was virtually impossible to do arithmetic without them, they were accorded the status of legitimate numbers. These three stages correspond more or less to the ancient, medieval, and recent periods, respectively. The Greeks (and probably the Babylonians and Egyptians as well) were well aware of negative numbers. However, they made little use of them, and they very explicitly denied that negatives could be numbers in the same sense as 1, 2, 3, . . . or fractions. Negative numbers were put to more extensive use during the latter half of the Middle Ages. Moslem and Christian mathematicians made considerable use of them, while simultaneously denying that they really were numbers. Near the close of the Middle Ages, Fibonacci pointed out that even though negative numbers are not really numbers they are of inestimable assistance in the task of recording financial transactions in which there are net losses. By the sixteenth century, the conflict between practice and belief had been resolved by accepting negatives as legitimate numbers. This allowed mathematicians of that century to work out the basic rules for adding, subtracting, multiplying, and dividing negative numbers. From their work, Pierre de Fermat (1601-1665) developed modern abstract arithmetic.

The development of complex numbers went through roughly the same three stages of initial discovery and denial, expanded usage accompanied by continued denial, and final acceptance. The earliest historical evidence of the discovery of complex numbers comes from ninth-century India. It is probable that Moslem mathematicians were acquainted with complex numbers. Complex numbers were not used with any great regularity until the sixteenth century. During the late sixteenth and early seventeenth centuries, they were extensively employed by European mathematicians. When Isaac Newton (1647-1727) and the philosopher Leibnitz(1646-1716) invented the calculus later in the same century, complex numbers became absolutely indispensible, however, their stature as legitimate numbers remained suspect. Finally, in 1799, a complete theory of the complex numbers was proposed by the Norwegian mathematician Wessel (1745-1818). Basically, Wessel showed how the algebra of complex numbers could be derived from classical geometry. From that point onward, the legitimacy of complex numbers was not in serious doubt. Although other examples of the broadening of the number concept could be given, the preceding two suffice to illustrate the general trend.

Concerning the other new types of numbers that evolved, it is sufficient to note that by the mid-1800s the number concept had been extended to include such things as irrational numbers, real numbers, hypercomplex numbers, algebraic numbers, and transcendental numbers. During the course of these somewhat haphazard extensions of the number concept into new territory, most mathematicians implicitly assumed that their new numbers were logically sound derivations from the original Greek number concept; that is, they assumed, as a matter of course, that these new numbers could be derived from the natural numbers. However, no one bothered to construct a credible mathematical demonstration of the truth of this assumption. During the second and third quarters of the nineteenth century, rigorous proofs of this sort became fashionable. In Germany, Leopold Kronecker (1823-1891) proposed that Pythagoras' conjecture that all mathematics is held hostage in the numbers 1, 2, 3, . . . is correct. "God made the integers," intoned Kronecker, "All else is the work of man." Following this dramatic assertion, Kronecker announced a "Pythagorean program" whose aim was to prove by rigorous methods that each of the new types of number evolved since the Middle Ages could be reduced to Pythagoras' natural numbers. About the same time and also in Germany, Richard Dedekind (1831-1916) and Karl Weierstrass (1815-1897) contributed work along the same lines as Kronecker's. Dedekind and Weierstrass proved that the most important of all the new types of number, the real numbers of algebra and analysis, can be derived in a step-wise manner from the natural numbers.

Thus, by the third quarter of the nineteenth century, the number concept had come full circle. First, there was Pythagoras' version of the concept which proved to be too narrow. This caused the concept to be extended, slowly at first and then very rapidly, to include many new forms of number. Finally, it was established to most mathematicians' satisfaction that these new numbers were not actually new: they all could be derived from 1, 2, 3. . . . After two and one-half mileniums, number lore had at last returned to its touchstone.

Third Phase

Having returned to the natural numbers, the question now became what was to be made of them. There seemed to be two clear alternatives. First, we simply could continue in the Pythagoras cum Kronecker tradition of accepting natural numbers as God-given and beyond the capacities of human understanding. Although this doctrine had been popular for 2,500 years, it did not square with the positivistic spirit that pervaded science and mathematics during the late nineteenth and early twentieth centuries. Second, we could maintain that, instead of being God-given, natural numbers were invented by human beings to fulfill certain necessary functions. During the fourth quarter of the nineteenth century this second alternative was advanced first by Gottlob Frege (1848-1925), then by Guiseppe Peano (1858-1932), and finally Bertrand Russell (1872-1969). It soon won widespread acceptance.

If we accept the suggestion that natural numbers were invented, then the question, "What are numbers?", can be interpreted in only one way. The question must be assigned the scientific interpretation we considered earlier in this chapter. The advent of the scientific interpretation was a momentous event in the history of number and it opened the door to entirely new realms of investigation. Unlike Pythagoras and Kronecker, we cannot stop with 1, 2, 3, . . . if we accept this interpretation. The mathematician is charged with attempting to construct formal definitions of the natural numbers in terms of more primitive ideas. The scientist who now enters the history of number lore for the first time is charged with tracing the growth of the natural number concept during the course of human cognitive development.

Throughout the remainder of this book we shall be concerned with how number lore has fared under the aegis of the scientific interpretation. In Part I, we shall be concerned primarily with mathematical contributions. We shall focus on attempts to define natural numbers in terms of two ideas borrowed from logic. We shall see that two distinct perspectives on how the natural numbers should be defined, an "ordinal" viewpoint and a "cardinal" viewpoint, emerged during the fourth quarter of the nineteenth century and the first decade of the twentieth. We also shall see that a third definitional perspective has emerged more recently. In Part II, we shall take up the psychological side of the scientific interpretation of "what are numbers?" We shall see that the definitional perspectives developed by mathematicians may be translated into psychological theories of how a child acquires the natural number concept during mental development. We shall review the principal findings of psychological studies of number development conducted since the 1920s. We shall see that psychological research suggests a close parallel between mathematical definitions of the natural number concept and the facts of number development in children's thinking.

PART I

THE LOGIC OF NUMBER

2

LOGICAL PRELIMINARIES

> But for the moment the natural numbers seem to represent what is easiest and most familiar in mathematics. But though familiar they are not understood. Very few people are prepared with a definition of what is meant by "number," "0," or "1."
>
> Bertrand Russell

As we saw in the preceding chapter, rigorous proofs of the conjecture that all of the various types of numbers employed in classical mathematics are held hostage in the system of natural numbers began to appear during the latter half of the nineteenth century. One very important effect of these proofs was to establish that, in consequence of the fact that arithmetic takes the natural numbers as its starting point, all of so-called "higher" mathematics (that is, that beyond arithmetic and algebra) can be reduced to arithmetic without any loss of meaning. We also saw that the rediscovered importance of the natural numbers soon led foundationally oriented mathematicians to attempt what had been regarded as sacrilege for 25 centuries—to construct a formal definition of the natural numbers in terms of more primitive ideas.

The structure of this and the succeeding three chapters owes much to my personal correspondence with Filmer S. C. Northrop, Sterling Professor of Philosophy (Emeritus), Yale University. Although I had written on the topics dealt with in these chapters prior to this correspondence, there were certain errors in my thinking which Professor Northrop recognized and communicated to me. Therefore, I should like to acknowledge my debt to him. Needless to say, Professor Northrop is not to be held accountable for such errors as remain.

It is the task of this and the next three chapters to review what modern mathematics and, more to the point, mathematical logic have had to say about the primitive notions which the natural numbers presuppose.

In this particular chapter, we shall deal with some preliminary distinctions of a purely logical sort which play important roles in the theories of number to be taken up in chapters 3, 4, and 5. These preliminary matters fall into three broad categories. First, we shall examine what a definition of the natural numbers must accomplish if the rest of classical mathematics is to follow from it. * This general question frequently is obscured in number treatises and appears to have been the source of considerable misunderstanding and unwarranted controversy. Second, a brief synopsis of modern mathematical logic will be offered. In this very selective sketch, three branches of logic—the abstract theory of propositions, the abstract theory of relations, and the abstract theory of classes—whose notions have been foci of formal definitions of the natural numbers, are emphasized. Finally, we shall consider some logical, or "metamathematical," criteria in virtue of which the adequacy of these formal definitions commonly is judged.

GOALS OF A NATURAL-NUMBER THEORY

To avoid confusion, it is important to be clear about precisely what one seeks to accomplish by exhibiting a definition of the number concept. Depending on one's scholarly point of view, a great many contrasting aims suggest themselves. Philosophers, like the early Pythagoreans, might seek a characterization of number which reflects their presuppositions about the nature of reality, the nature of human knowledge, and so forth. Experimental psychologists might seek a characterization which reflects what they know about the manner in which numbers routinely are employed in thinking, problem solving, and everyday discourse.

Insofar as mathematics is concerned, the principal objective of any formal definition of number must be to make mathematics itself possible. That is, the entire body of classical mathematics must follow from such a definition. On the whole, this may seem a rather herculean aim. However, we already have seen that the scope of this

---

*By "classical mathematics," we shall understand the entire body of mathematics as of about 1900. We shall explicitly exclude from this notion such recent developments as topology, set theory, and mathematical logic.

problem can be narrowed to manageable proportions. For example, we know, thanks to the work of Dedekind, Weierstrass, and others, that classical mathematics springs from arithmetic and, hence, from the natural numbers. Therefore, if our aim is a definition which makes possible the larger body of mathematics, it suffices to construct a definition from which the theorems of arithmetic can be derived. But the problem is even narrower than this.

At one time or another, usually in junior high or high school, most of us were taught that all the theorems of arithmetic follow from just three simple principles, which are usually termed the "first laws" of arithmetic. These first laws are:

1. Commutativity: The result of adding or multiplying two natural numbers, a and b, is independent of the order in which they are combined. That is, $a + b = b + a$ and $ab = ba$.

2. Associativity: The result of adding or multiplying three natural numbers, a, b, and c, is independent of the order in which they are combined. That is, $(a + b) + c = a + (b + c)$ and $(ab)c = a(bc)$.

3. Distributivity: The result of adding two natural numbers, a and b, and multiplying by a third natural number, c, is the same as separately multiplying the first two numbers by the third and adding the products. That is, $(a + b)c = ac + bc$.

Thus, our definitional problem comes down to these first laws of arithmetic. In order to make mathematics as a whole possible, it is sufficient to construct a definition of the natural numbers which makes commutativity, associativity, and distributivity possible.

A second mathematical goal of any formal definition of number concerns the consistency of mathematics. Mathematical consistency generally means that if the truth of some particular statement or theorem can be proven, then it follows that its converse cannot be proven. If arithmetic is consistent, for example, then the statement "2 is an even number" can be shown to be true or false but not both. It is known that the consistency of mathematics cannot be proved without a prior reduction of the number concept to more primitive notions (see, for example, Brainerd 1973a, Stabler 1953). This is so because when the number concept is left undefined it becomes our most basic level of analysis. Under such conditions, a definitive proof of the consistency of mathematics is not possible because, as part of the demonstration, it must be assumed without proof that the system of natural numbers is itself consistent. In the absence of a consistency proof, mathematicians must go about their work without any assurance that the ends which they pursue (the generation of consistent and comprehensive theories) are theoretically attainable.

Thus, a formal definition of the number concept is a necessary precondition for any demonstration of the consistency of mathematics.

This is not to say that such a demonstration automatically follows from an adequate definition. On the contrary, two celebrated theorems, proved more than four decades ago by Gödel (1930, 1931), established certain unforeseen obstacles to such a demonstration. Explicitly, Gödel showed that, above the level of arithmetic, an indeterminacy or trade-off obtains between consistency and completeness such that neither consistency nor completeness can be established without sacrificing the other. But these results presuppose a more thorough grasp of the notions of "consistency" and "completeness."

Therefore, we must, at a minimum, formulate a definition which makes possible the first laws of arithmetic and, consequently, the entire body of classical mathematics. Generally speaking, this suggests that we search for some primitive idea or ideas (1) which can be shown to be presupposed by the natural numbers and (2) on which the first laws of arithmetic can be shown to turn. But where are such ideas to be found? Clearly, they cannot come from mathematics, where natural numbers are the most basic level of analysis. Since the rest of mathematics is defined in terms of the natural numbers, it is obvious that mathematics cannot define the natural numbers. During the last two decades of the nineteenth century, Frege, Peano, and especially Russell argued persuasively that we should look to logic as a source of the primitive ideas in virtue of which the number concept may be defined. According to their "logicist thesis," mathematics in general and the natural numbers in particular can be derived from the concepts of logic. Although this view is expressed in both Frege's _Die grundlagen der arithmetik_ (1884) and Peano's _Formulaire de mathématiques_ (1894-1908), Russell's _Principles of mathematics_ (1903) is widely regarded as the definitive formulation of the thesis. Later, Whitehead and Russell provided a formal deduction of mathematics from logic in what is perhaps the most celebrated mathematical work of the twentieth century, _Principia mathematica_ (1910-1913). Before we can grasp the significance of the logicist thesis, however, we must understand precisely what is meant by the term "logic."

LOGIC

Logic has one common-sense meaning and two technical meanings. In everyday parlance, as well as in the language of most scientists, logic refers to the nature of argument. If a certain conclusion follows ineluctably from certain premises which are either assumed (in the case of laypersons), or established experimentally (in the case of the scientist), then the train of argument from premises to conclusion is declared "logical." One technical meaning of logic refers to formal logic, which is connected with the study of the syllogism. (The classic illustrative syllogism: All persons are mortal. Socrates is a person. Therefore, Socrates is mortal.) The other technical meaning refers to modern mathematical logic. It is with the latter two meanings of logic that we shall be concerned here.

Formal logic has been with us for roughly 23 centuries, and it generally is believed to have begun with Aristotle. Mathematical logic, on the other hand, is only slightly more than a century old. By mutual agreement, the publication of George Boole's An Investigation of the Laws of Thought (1854) is regarded as the birth of mathematical logic. Augustine De Morgan, an Englishman, and Charles Pierce, an American, also were influential in its early development. It was not until the final decade of the nineteenth century and the first decade of the twentieth that mathematical logic became firmly established as a new field of study. The credit for establishing mathematical logic as a viable entity goes primarily to Guiseppe Peano, Bertrand Russell, and Alfred North Whitehead, especially the latter two.

In view of its Aristotlean origins, formal logic was associated historically with the study of Greek. Throughout the course of its long evolution, its connections with mathematics were few and, for the most part, coincidental. Whereas theologians and philosophers typically were well versed in formal logic—it will be recalled that "logic chopping" was a favorite pastime of medieval scholars—mathematicians, almost without exception, were not. Mathematical logic, in contrast, always has been closely affiliated with the study of mathematics. Its principal founders were either mathematicians or mathematical philosophers by training. Moreover, one of the abiding themes of mathematical logic has been the construction of definitive statements about the nature of mathematical proof, what proofs tell us, and what criteria should be used to evaluate proofs (see, for example, De Long 1970, Quine 1951). Despite the difference in the degree of their historical connection with mathematics, formal logic and mathematical logic deal with the same general subject matter—the structure of valid deductive reasoning. That is, both are concerned with processes which underlie the formulation of statements of the general type "A implies B." Both are concerned with what it means to say that such statements are true and, consequently, with the nature of proof. Given the same subject matter, there are technical differences between the methods that formal logic and mathematical logic employ to study this subject matter. Not surprisingly, the methodology of mathematical logic bears a more pronounced resemblance to the methodology of mathematics itself, particularly the methodology of classical geometry, than does the methodology of formal logic. Further, and this is why mathematical logic has almost completely supplanted formal logic today, these methods are considerably more powerful than those of formal logic in that they are capable of proving or disproving a much greater range of propositions.

Perhaps the most significant truth discovered by formal logic is that the validity of deductive arguments turns upon their structure rather than, as common sense would suppose, upon the truth or the meaning of their premises. For example, recall our illustrative argument; "All persons are mortal. Socrates is a person. Therefore, Socrates is mortal." This argument is valid not because it so happens that humans die and Socrates is of human kind. Its validity is a consequence of the fact that it is a special case of a generic type of argument whose members all have the following structure:

(1) For all Cs, x is true;
(2) y is a member of C;
(3) Therefore, x is true of y.

Regardless of the meaning of C, x, and y, (1) through (3) is a valid argument.

It was the aim of formal logic to formulate and catalogue exhaustively those argument structures which produce valid conclusions. In this, formal logic failed. Its failure usually is ascribed to formal logicians' interest in making normative statements about the use of deductive arguments in areas such as jurisprudence and theology. Formal logicians, according to this interpretation, failed to produce a sufficiently exhaustive formulation of logic because they were not concerned with logical knowledge—or rather knowledge about logic—as an end in itself. Whatever the reasons, the deficiencies of formal logic were all too apparent by the mid-nineteenth century. Among other things, it was widely recognized by that time that there are some very important argument forms whose validity seems incontestable that formal logic does not encompass. Most of these argument forms come from mathematics and, hence, involve premises and conclusions which are quantitative. One elementary example is "The square root of 5 is greater than the cube root of 3. The square root of 7 is greater than the square root of 5. Therefore, the square root of 7 is greater than the cube root of 3." This argument is obviously valid, and its validity turns on the fact that it is a member of a class of arguments of the following form:

(4) $p > q$
(5) $r > p$
(6) therefore, $r > q$.

Although (4) through (6) is a valid argument whatever the respective meanings of p, q, and r, it cannot be formulated in the language of formal logic (see Cohen and Nagel 1934).

Mathematical logic originated as a means of extending logic to cover the study of argument forms such as (4) through (6). Since the turn of the century, however, the principal impetus for the development of mathematical logic has been the logicist thesis. An important consequence of this work has been the demonstration beyond reasonable doubt that, whatever else it may be concerned with, mathematics treats a small number of logical ideas. However, this statement should not be taken to imply that everyone is satisfied that mathematics is nothing more than logic. There has been much acrimonious debate on this point between the proponents of two schools of mathematical philosophy: logicism and intuitionism. Logicism, founded by Bertrand Russell and kept in good repair today by W. V. O. Quine, maintains that a complete reduction of mathematics to logic is possible and that, in fact, it already has been effected. It is argued that Principia mathematica and subsequent work establish that all mathematics flows from exactly three primitive logical concepts which are roughly analogous to the English terms "all," "is," and "neither/nor." (The third term frequently may be replaced by "not both.") To readers who are familiar with the complexities of mathematics but unfamiliar with mathematical logic, this claim no doubt seems difficult to countenance. It also seems unreasonable to mathematicians of the intuitionist school. Intuitionism was founded by L. E. J. Brower, a Dutch mathematician, a few years after Russell's Principles of Mathematics was published. According to intuitionism, some content is irretrievably lost when mathematics is translated into the language of mathematical logic. In place of the logicist thesis, intuitionists adopt a modified version of the Pythagorean hypothesis that natural numbers form the ultimate and irreducible foundation of mathematics.

To avoid obscuring our main theme any more than is necessary, the question of whether or not a complete reduction of mathematics to logic is possible will be taken as currently indeterminate and better left to philosophers to agonize over. The indeterminacy of the question notwithstanding, twentieth-century mathematical logicians have shown this much with comparative certainty. Mathematics deals with rudimentary logical concepts. Whether or not this is absolutely all that mathematics deals with is quite irrelevant for our purposes. For us, the single fact that the primitive ideas of logic are in some sense involved in mathematics is sufficient to suggest that these ideas might be a very fruitful source of hypotheses about the meaning of key mathematical concepts such as number.

This brings us back to the problem of defining the system of natural numbers in terms of the more fundamental ideas which the system presupposes. From the preceding, it should be apparent that mathematical logic probably is a good place to look for such ideas. But precisely where within mathematical logic are these ideas likely to be found? To answer this question, we must know more about the subject matter of mathematical logic. At the turn of the century, mathematical logic was exceedingly amorphous. Its boundaries were not sharply defined or even intuitively understood. Russell set out to rectify this situation. In his Principles of Mathematics and a series of papers which preceded it, he presented some preliminary views on precisely what mathematical logic is concerned with. A mathematically rigorous development of these views subsequently was presented in Principia mathematica. From our perspective, the key aspect of Russell's analysis of the subject matter of mathematical logic is that he divided it into three branches: (1) the study of statements about statements (propositional logic), (2) the study of statements about relations (logic of relations), and (3) the study of statements about classes (logic of classes). Although post-Principia work in mathematical logic has wrought many changes in Russell's early formulations, his distinction between propositional logic, the logic of relations, and the logic of classes remains fundamental (see Quine 1951). Moreover, this distinction provides a very convenient framework within which to explicate the logical foundations of the number concept.

Of the three branches of mathematical logic, the logic of propositions is the most fundamental. Its primitive ideas are employed in both of the other two branches but not conversely. Individual statements about either relations or classes always may be construed as propositions. Moreover, the concept of relation and the concept of class both may be defined as special cases of a propositional concept to be considered shortly—propositional function. Because propositional logic is more basic than the other two branches, it also is further removed from the language of mathematics in general and number in particular. Importantly, propositional logic is concerned with purely categorical statements—that is, those which may be taken to be either clearly true or clearly false. Consequently, the language of propositional logic does not permit one to formulate the familiar quantitative statements, for example, of arithmetic. On the other hand, such statements appear to be possible in the language of both the logic of relations and the logic of classes. Hence, insofar as we are concerned to understand the logical roots of the number concept, we shall have to examine these two branches of logic in greater detail. Fortunately, we do not require an exhaustive treatment of either. For our purposes, it will be quite sufficient to review only those relational and classificatory distinctions which are central to the number doctrines taken up in Chapters 3, 4, and 5. Before doing so, however, perspicuity demands that we briefly consider some concepts of propositional logic which are involved in making statements about relations and classes.

The Logic of Propositions

Not surprisingly, the starting point of propositional logic is the concept of proposition itself, which is accepted as an undefinable notion. In lieu of saying nothing at all about the concept, "proposition" is intuitively characterized as "anything that is true or that is false" (Russell 1903: 12). In addition to this somewhat vague description, it is assumed that our informal psychological knowledge of the concept of proposition is sufficiently valid that we shall not be led into gross error without a formal definition. As noted earlier, propositional logic is concerned with categorical statements. Propositions must be capable of being either unequivocally true or unequivocally false and nothing in between. "Rutherford is a Nobel laureate," "Pascal is a mathematician," and "the sum of the interior angles of a plane triangle is 180 degrees" are all propositions in this sense. However, "it is raining" is not. If I look out the window to verify the latter statement, I may observe that it is misting, a state which lies somewhere between raining and not raining. Propositions also must be what is called "particular" in formal logic. That is, they may not contain quantifying predicates such as "all," "some," "no," or "always." Thus, "all Americans are rich," "no member of Congress is a mathematician," and "some Canadians are hockey players" are not propositions in this sense.

The concept of truth, like the concept of proposition, remains undefined in propositional logic. We again assume that our informal psychological knowledge of the concept is sufficiently valid that we shall not commit errors. Moreover, we now know that this assumption is absolutely inescapable. One of the most spectacular achievements of post-Principia logic has been the discovery that a formal syntactic definition of "truth" simply cannot be given within the confines of any logical system in which the definition is to be used. Alfred Tarski (1939) constructed the first formal proof of this somewhat surprising fact.

In the case of most common propositions, their truth is an empirical rather than a logical question. Therefore, propositional logic is not concerned to ascertain the truth of individual propositions. That is, whereas logical truth depends only on the structure of a train and not on the content of its propositions, propositional logic is concerned, instead, with showing how the truth of "trains" of connected propositions turns on the truth of the individual propositions in the train, especially when these propositions contain so-called "propositional variables." For example, consider the train, "If Newton was a physicist, then Columbus discovered America." If "Newton was a physicist" happens to be false, then the train is true. Second, if "Newton was a physicist" happens to be true, then the train is true only if "Columbus discovered America" also is true. More generally, if the propositional variables p and q denote any two proposition whatsoever, then "if p then q" is true if p is false, and it also is true if p and q both are true.

To formulate exhaustively the ways in which trains of two or more propositions turn on the truth values of their constitutent propositions, it is necessary to formulate the possible connections of two or more propositions. Given any two propositions, p and q, which may be either true or false, there are precisely 16 possible connections or composite propositions. All 16 composites are presented in Table 2.1. The name and standard notation for each are given in the left-hand column. This table is the core of propositional logic. Although it contains only the 16 possible connections of two propositions, any larger composite proposition involving more than two propositions may be expressed in terms of these 16 binary composites. Of these 16 composites, number 5 (nonconjunction) and number 15 (nondisjunction) are the most important. Given either of these, the remaining 15 composites can be defined.

Certain composite propositions are always true regardless of the truth values of their constituent propositions. These are called "valid formulas" and they are similar in principle to the theorems of mathematical systems such as Euclidean geometry or arithmetic. All of the valid formulas of propositional logic can be shown to be made up entirely of one or more of three formulas. These are called "first laws" and they are to propositional logic what the first laws of arithmetic are to arithmetic. Expressed in the notation introduced in Table 2.1, the three first laws are:

1. Law of Affirmation of the Consequent: Given any two propositions p and q, the composite $p \rightarrow (q \rightarrow p)$ is always true. This law translates as: If p, then 'if q then p'.

2. Distributive Law of Implication: Given any three propositions p, q, and r, the composite $[p \rightarrow (q \rightarrow r)] \rightarrow [(p \rightarrow q) \rightarrow (p \rightarrow r)]$ is always true. This law translates as: If "if p then 'if q then r'," then "if 'if p then q' then 'if p then r'."

3. Converse Law of the Contrapositive: Given any two propositions p and q, the composite $(\text{not-}p \rightarrow \text{not-}q) \rightarrow (q \rightarrow p)$ is always true. This law translates as: If "if not-p then not-q" then "if p then q."

Finally, we must consider the very important concept of "propositional function." Propositional functions are statements otherwise resembling propositions which contain one or more indeterminate terms. These indeterminate terms may take on any number of values, For example, the four statements on the left are propositions and their counterparts on the right are propositional functions:

TABLE 2.1

The 16 Outcomes from Combining Two Binary Propositions

Outcome	Component propositions			
	p = true q = true	p = true q = false	p = false q = true	p = false q = false
Affirmation (p * q)	true	true	true	true
Disjunction (p ∨ q)	true	true	true	false
Reverse Implication (p ← q)	true	true	false	true
Implication (p → q)	true	false	true	true
Nonconjuction (p/q)	false	true	true	true
Affirmation of p (p)	true	true	false	false
Affirmation of q (q)	false	false	true	true
Equivalence (p ≡ q)	true	false	false	true
Denial of q (~q)	false	true	false	true
Denial of p (~p)	false	false	true	true
Nonequivalence [~(p ≡ q)]	false	true	true	false
Conjunction (p · q)	true	false	false	false
Nonimplication [~(p → q)]	false	true	false	false
Nonreverse implication [~(p ← q)]	false	false	true	false
Nondisjunction [~(p ∨ q)]	false	false	false	true
Negation [~(p * q)]	false	false	false	false

Source: Compiled by the author.

(1) Verdi composed operas.	(1') a composed operas.
(2) Edison invented appliances.	(2') a invented appliances.
(3) The British Royal Society was founded in 1662.	(3') a was founded in 1662.
(4) Lincoln was assassinated.	(4') a was assassinated.

The important thing to note about these illustrations is that the truth values of the propositions on the left are clear but the truth values of the propositions on the right are not. For example, statement (4) is clearly true. However, statement (4') is true for some values (McKinley and Kennedy), false for other values (Roosevelt and Eisenhower), and meaningless for other values (triangle and street car). A propositional function may be defined more precisely as follows: A propositional function is any statement containing one or more variable terms which becomes a proposition upon assigning these terms suitable values. The range of suitable values of a variable term is called its "domain." Within a domain, those values which produce a true proposition are said to satisfy the propositional function.

Subscript notation usually is employed to distinguish propositions and propositional functions. Hence, if p is any proposition, p_a is a propositional function containing one variable term, p_{ab} is a propositional function containing two variable terms, p_{abc} is a propositional function containing three variable terms, and so on. More important, everything that has been said up to this point about propositions is true also of propositional functions.

The Logic of Relations

Relations are concerned with propositional functions in two unknowns. Explicitly, a relation is anything that makes a propositional function containing two variable terms (for example, "a is a citizen of b") true. Given any propositional function of the form, R_{ab}, where A is the domain of a and B is the domain of b, we may say that R_{ab} "relates" some value a_i from A and some value b_j from B if and a_i and b_j satisfy R_{ab}. Thus, "is the wife of" may be said to relate "Josephine" and "Napoleon" but not "Helen" and "Achilles" because a = Josephine and b = Napolean converts "a is the wife of b" to a true proposition, but a = Helen and b = Achilles does not.

The subject matter of the logic of relations is primarily taxonomical. That is, it is concerned to say what general sorts of empirical, logical, and mathematical relations there are. Consequently, the logic of relations seeks to isolate those characteristics of relations in virtue of which they may be said to differ. These characteristics are called the "formal properties" of relations. For purposes of explicating the number concept, we need to be familiar with three of these formal properties: symmetry, transitivity, and reflexivity.

Symmetry

We have a function, R_{ab}, and two values, a = x and b = y, which satisfy R_{ab}. We say that R_{ab} is symmetrical if, for all values x and y that satisfy R_{ab}, it is the case that R_{xy} and R_{yx} both are true. We say that R_{ab} is asymmetrical if, for all values x and y that satisfy R_{ab}, it is the case that R_{yx} is false when R_{xy} is true and, conversely, R_{xy} is false when R_{yx} is true. Finally, we say that R_{ab} is nonsymmetrical if R_{yx} is sometimes true when R_{xy} is true. Brotherhood is a common symmetrical relation. If x is the brother of y, then it follows that y also is the brother of x. Parenthood, on the other hand, is an equally common asymmetrical relation. If x is the father or mother of y, then y cannot be the father or mother of x. Relations of preference are common instances of nonsymmetry. For example, if philosophers admire scientists, it may or may not be the case that scientists admire philosophers. Finally, concerning the relations which obtain among numbers, these relations are either symmetrical ("equal to"), asymmetrical ("greater than"), or nonsymmetrical ("less than or equal to").

Transitivity

We have a function, R_{ab}, and three values a = x, y and b = y, z, which satisfy R_{ab}. It is known that R_{xy} and R_{yz} both are true propositions. We say that R_{ab} is transitive if R_{xz} must be true when R_{xy} and R_{yz} are true. We say that R_{ab} is intransitive if R_{xz} must be false when R_{xy} and R_{yz} are true. We say that R_{ab} is nontransitive if R_{xz} may be either true or false when R_{xy} and R_{yz} are both true. Symmetrical relations are always transitive. If x is the brother of y and y is the brother of z, then x and z also must be brothers. Some asymmetrical relations are intransitive. If x is the father of y and y is the father of z, then x cannot be the father of z. Other asymmetrical relations are transitive. If x is older than y and y is older than z, then x also is older than z. Hence, the earlier symmetry-asymmetry-nonsymmetry distinction cannot be reduced to the present transitivity-intransitivity-nontransitivity distinction. Relations of preference serve to illustrate nontransitivity. If theologians admire philosophers and philosophers admire scientists, it may or may not be true that theologians admire scientists. Finally, turning to the common relations among numbers, some are transitive ("equal to," "greater than"), some are intransitive ("natural log of"), and some nontransitive ("square root of").

Reflexivity

We have a function, R_{ab}, and two values a = x and b = y, which satisfy R_{ab}. We say that R_{ab} is reflexive if, for all values x and y that satisfy R_{ab}, it is the case that R_{xx} and R_{yy} are true. We say that R_{ab} is irreflexive if, for all values x and y that satisfy R_{ab}, it is the case that R_{xx} and R_{yy} both are false when R_{xy} is true. We say that R_{ab} is nonreflexive if, for all values x and y that satisfy R_{ab}, it is the case that R_{xx} and R_{yy} may or may not be true when R_{xy} is true. The symmetrical relation of personal identity serves to illustrate reflexivity. If it is true that Sir Walter Scott is the same person as the author of Waverly, then "Sir Walter Scott is the same person as Sir Walter Scott" and "the author of Waverly is the same person as the author of Waverly" are true statements also. Other symmetrical relations (for example, brotherhood) are not reflexive. Asymmetrical relations always are irreflexive. For all values x and y that make "x is the father of y" a true proposition, the propositions "x is the father of x" and "y is the father of y" are false. Preference relations again may be used to illustrate nonreflexivity. If philosophers admire scientists, it may or may not be true that philosophers and scientists admire themselves. Finally, the common relations among numbers include examples of reflexivity ("equal to"), irreflexivity ("greater than"), and nonreflexivity ("less than or equal to").

The Logic of Classes

We turn now to the third and last topic in our brief survey of mathematical logic—the logic of classes. Classes, like relations, are concerned with propositional functions. But classes are concerned with propositional functions with two variable terms. A class is anything which makes a propositional function in two unknowns true. To illustrate, consider the proposition, "Henry James is the author of The Golden Bowl." If we substitute the variable term 'a' for 'Henry James' and the variable term 'b' for 'The Golden Bowl', then we have constructed a function R_{ab} that is satisfied by all values of a and b such that a is some author and b is some work that he or she has written. In contrast, by leaving 'Henry James' in the above proposition and substituting the variable term 'a' for 'The Golden Bowl', we have defined the class of works of which Henry James is the author. More generally, the concept of class may be defined as follows: Given some propositional function R_n containing one variable term n, all those values from the domain of n which produce a true proposition when substituted for n in R_n are said to be a "class" (or "set"). The propositional function R_n is said to "determine" the class of values that satisfy it.

The notion of membership is a pivotal concept in the logic of classes. Given some propositional function R_n, those values of n that satisfy the function are called "members" or "elements" of the class which the function determines. Given some class C and one of its members x, membership usually is symbolized $x \in C$. It is important to note that membership clearly is a relational concept—i. e., it is a propositional function in two unknowns of the general form "m belongs to n." Membership may be distinguished from the other relations discussed above principally by the fact that the second variable term in the propositional function is a propositional function. It is important to note that there is nothing in the definition of relation considered earlier that would tend to exclude this possibility. Thus, any relational statement of the form, "m is a member of R_n," is a statement of membership, and it may be denoted $n \in R_n$. Although membership was not discussed in the preceding section on relations, the taxonomy of formal properties developed there applies to membership also. Explicitly, membership can be shown to be a symmetrical, nontransitive, and reflexive relation. Demonstration of this fact is left to the reader (see Russell 1903 for a detailed treatment).

We saw that the logic of relations is concerned with stating what the formal properties of relations are. Similarly, the logic of classes is concerned to formulate the formal properties of classes. The two most important of these are "intension" and "extension." The intension of a class is a quality which is shared by all members of the class in virtue of their membership in the class. The members of the class of poets share the property of being poetic, the members of the class of triangles share the property of being triangular, the members of the class of pedagogues share the property of being pedagogical, and so on. As these examples suggest, the intension of a class is synonymous with the meaning of the propositional function which determines the class. In contrast, the extension of a class is synonymous with the particular values that satisfy the defining function. The intension of a class is logically more fundamental than its extension: Classes with the same intension perforce must have the same extension, but classes with the same extension may have quite different intensions. For example, "x is a man" and "x is a featherless biped" are very different intensions which happen to have precisely the same extension. Classes that have the same extension are termed "equal." Classes that have the same intension are termed "identical."

In addition to membership—the relation between a class and its elements—the theory of classes treats another relational notion which is fundamental to the theory of number discussed in Chapter 4. This notion is variously called "correlation" or, more commonly, 'correspondence." Generally speaking, correspondence is concerned with the relative manyness or populousness of the elements of classes.* Correspondence, like membership, may be precisely defined by appealing to the concept of propositional function. Suppose there are two propositional functions, R_a and R_b, and two classes, A and B, determined by R_a and R_b respectively. Correspondence is concerned with how many elements are members of the two classes—how many values satisfy the respective propositional functions. Normally, relative manyness is determined by connections among the elements of the respective classes which are implicit in the propositional functions. If for every $a_i \in A$ which satisfies R_a there corresponds one and only one $b_j \in B$ which satisfies R_b and conversely, then we say that the correspondence between A and B is "one to one." If R_a = "x is a monogamously wedded male" and R_b = "x is a monogamously wedded female," then the correspondence between the two classes defined by these functions is one to one. If for every $a_i \in A$ which satisfies R_a there corresponds more than one $b_j \in B$ which satisfies R_b, and for every $b_j \in B$ which satisfies R_b there corresponds one and only one $a_i \in A$ which satisfies R_a, then we say that the correspondence between A and B is "one to many." If R_a = "x is a polygamously wedded male" and R_b = "x is a polygamously wedded female" then we have a one-to-many correspondence. The converse of this correspondence is many-to-one correspondence, which may be illustrated by R_a = "x is a polyandrously wedded male" and R_b = "x is a polyandrously wedded female." Finally, if for every $a_i \in A$ which satisfies R_a there corresponds more than one $b_j \in B$ which satisfies R_b and conversely, then we say that the correspondence between A and B is many to many. If R_a = "x is a bigamously wedded male" and R_b = "x is a bigamously wedded female," then we have a many-to-many correspondence.

The notion of a one-to-one correspondence between two classes frequently is confused with the notion of equality or "same extension." Hence, it is worth emphasizing, in closing, that the mere fact that there is a one-to-one correspondence between the elements of two classes does not entail that they have the same extension. If R_a = "x is a human being" and R_b = "x is a featherless biped," then it so happens that the correspondence between the classes determined by these

---

*In the elementary grades, especially during recent years, most North Americans are taught to view number as synonymous with manyness. Hence, many readers undoubtedly will be tempted to substitute "number" in this sentence. This would be a mistake, although a perfectly understandable one, that could lead to subsequent confusion. We shall see later that it is not at all clear to those who have considered the problem that number and manyness are the same thing.

functions is one to one and the classes have the same extension. However, if R_a = "x is an apostle of Christ" and R_b = "x is a month of the year," then the correspondence between the two classes is still one to one but their extensions consist of very different elements.

CRITERIA FOR MATHEMATICAL THEORIES

Our final task in this chapter is to consider the criteria in virtue of which mathematical theories are said to be acceptable or unacceptable. These criteria are variously called "logical," "metalogical," and "metamathematical." What these different adjectives are supposed to convey is that the criteria we shall be discussing are not a part of mathematics proper. They lie somewhere outside it. In view of the fact that these criteria are used to evaluate mathematics itself, it is obvious to common sense that they must come from somewhere other than mathematics. Otherwise attempts to evaluate any portion of mathematics would be doomed to circularity: In order to demonstrate the acceptability of some portion of mathematics, we first would have to assume the acceptability of some other aspect of mathematics (the criteria).

Although there are others, the three principal criteria by which mathematical theories are judged are consistency, completeness, and categoricalness. Given any mathematical theory, we usually ask, It is consistent? Is it complete? Is it categorical? This also is the order of importance of the three questions. Consistency is by far the most important of the three. Mathematical theories that are known to be inconsistent are of no interest whatsoever; on the other hand, incomplete and noncategorical theories may be of considerable interest. A mathematical theory is said to be consistent if, under all possible interpretations of its undefined concepts, it can be shown that the theory leads to no contradictory statements. In other words, it can be shown that for every statement it is not possible to prove both the statement and its converse. This characterization of mathematical consistency obviously squares with our common sense understanding of consistency. But consistency also has a somewhat more technical meaning.

Following the example of geometry, a mathematical theory, where possible, is formalized in terms of a small number of axioms, which are known to be independent of each other, and one or more rules of inference, which are the only permissible ways in which the axioms may be connected to each other. We proceed to prove theorems by using these rules of inference to connect the axioms in various ways. Any statement which can be shown to be entirely the result of repetitive connections of the axioms via the rules of inference is said to be a theorem of the system. Technically speaking, consistency has to do with the rules of inference. In classical mathematics, axioms usually are statements whose validity appears obvious. Ideally, we would like the validity of theorems deduced from these axioms to depend only on the validity of the axioms themselves. In other words, under the assumption that the axioms are valid, the theorems should be valid also. We say that a mathematical theory is consistent if the truth of this statement can be proven. Now, the only way a system can possibly be consistent is if the rules of inference pass on or transmit the property of validity from the axioms to the theorems. Obviously, if (1) the axioms are assumed to be valid, and (2) the rules of inference are the only method for connecting the axioms to generate theorems, then the theorems can only be valid if the rules of inference transmit validity from the axioms to the theorems. We shall consider a theory in Chapter 3 to which this technical meaning of consistency may be applied, that is, Peano's formalization of arithmetic. Peano's arithmetic consists of four axioms and one rule of inference (finite mathematical induction). Assuming the validity of Peano's axioms, we shall want to know whether or not the rule of finite mathematical induction transmits validity from the axioms to the theorems of arithmetic.

Following consistency in overall importance is completeness. We shall speak of two basic forms of completeness: "syntactical" (or "structural") completeness and "semantical" (or "functional") completeness. To explicate these ideas, lets us again assume that we have some hypothetical mathematical system which, like Euclidean geometry, has been formalized in terms of certain axioms and certain rules of inference. Syntactical completeness, like consistency, is concerned with the transmission of validity. Recall that consistency is concerned with whether or not every theorem generated by using the rules of inference to connect the axioms must be valid if the axioms are valid. Syntactical completeness is almost exactly the reverse: Can it be shown that every valid statement which is not an axiom also is a theorem of the system? Simply put, consistency is concerned with whether or not every theorem of the system is valid (assuming the axioms are valid), whereas completeness is concerned with whether or not every valid statement is a theorem. Semantical completeness on the other hand, focuses on the "expressive power" of our hypothetical system. Can the expressive power of our system be improved by adding further axioms, rules of inference, or both, or can we say all that it is possible to say with the axioms and rules of inference at hand? Thus, semantical completeness is basically concerned with the richness of the language provided by the axioms and rules of inference. If new theorems can be generated by adding further axioms or rules of inference, then our system is not semantically complete. However, if it can be shown that new theorems do not accrue from adding axioms, rules of inference, or both, then we say that our system is semantically complete.

The distinction between syntactical and semantical completeness has become important largely as a result of Kurt Gödel's widely acclaimed studies of consistency and syntactical completeness (Gödel 1930, 1931). To many readers, it probably seems intuitively obvious that syntactical completeness and consistency, being more or less opposite sides of the same coin, should follow from each other. It turns out that this is not so. Not only do consistency and syntactical completeness fail to imply each other, Gödel demonstrated that a very disconcerting indeterminacy obtains between the two in classical mathematics. Gödel showed that in every mathematical system sufficiently rich to include arithmetic (all of classical mathematics), it is impossible to demonstrate both the consistency and syntactical completeness of the system using only the methods of that system. Consistency can be demonstrated only at the expense of syntactical completeness and syntactical completeness can be demonstrated only at the expense of consistency (see Nagel and Newman 1968). However, Gödel's celebrated work does not exclude the possibility that a system sufficiently rich to include arithmetic can be shown to be both consistent and semantically complete. From our standpoint, it will be much more important to know that we can say all that we wish to say without inconsistency than it will be to know that we can demonstrate the "theoremhood" of everything that we say.

Finally, we come to categoricalness. Categoricalness is broadly concerned with the extent to which a mathematical theory may be said to treat a narrowly delimited class of phenomena—that is, a single propositional function of the form P_a. The meaning of this somewhat abstract statement is rather simple. Again, suppose we have some mathematical theory consisting of certain axioms and rules of inference. Now, as we shall see in Chapter 3 with Peano's arithmetic, axioms are composed of propositional functions. Because axioms consist of propositional functions rather than propositions, they do not tell us precisely what values are capable of satisfying them. Moreover, since different axioms always consist of different propositional functions (they are "independent" of each other), different values may be capable of satisfying different axioms. However, if we can find a certain class of values determined by a propositional function P_a that converts all of the axioms into true propositions, then we say that this class is an "interpretation" of our theory. Categoricalness refers to how many interpretations of a theory there are. We say that a theory is categorical when there is only one interpretation which satisfies the axioms. Categoricalness also may be thought of in terms of the previously discussed notion of class extension. If a mathematical theory is categorical, then all of the various classes whose members convert the axioms into true propositions have exactly the same extension. Categorical theories have the obvious characteristic of being concrete. We are about to see that concreteness has been viewed as an important virtue in theories of number.

3

THE RELATIONAL VIEW OF NUMBERS

> I regard the whole of arithmetic as a necessary, or at least natural, consequence of the simplest arithmetic act, that of counting, and counting itself as nothing else than the successive creation of the infinite series of positive integers in which each individual is defined by the one immediately preceding.
>
> Richard Dedekind

This chapter and the two which immediately follow it explore logical doctrines about the nature and meaning of the number concept. In all three of these chapters, we shall follow more or less the same course. Our first task, in each case, will be to review the key facts about some particular doctrine. Next, we shall consider the major objections which either have, or might have, been lodged by proponents of other viewpoints. Finally, we shall try to arrive at a balanced assessment of just how much damage is done to each theory by the various objections.

As mathematics goes, most of what is said in this and the following two chapters is quite elementary. However, much of it is technical in nature and, consequently, may try the patience of nonmathematical readers who have not previously encountered such material elsewhere. For these readers, a concrete preliminary characterization may serve to encourage perseverance in some of the denser passages. Generally speaking, the theories which will concern us are derived from one or both of two reasonably simple ideas: (1) identifying the natural numbers with ordinal numbers, and (2) identifying the natural numbers with cardinal numbers. Therefore, for the sake of perspicuity, let us first say, briefly, what ordinal and cardinal numbers are.

Suppose there is some collection of terms (x_1, x_2, x_3, . . .), and suppose that these terms are arranged in a fixed order of some sort. By a fixed order, we mean simply that there is a "first" term, a "second" term, and so on. Any such collection of discrete terms is called a "progression" or "series." When a progression has a first term, no last term, and no repeated terms, it is called a "simple infinite progression." When a progression has both a first term and a last term, it is called a finite progression. The series 1, 2, 3, . . . obviously can be used to represent the respective terms in any finite or simple infinite progression. The first term is called "1," the second term is called "2," and so on. Whenever the natural numbers are used in this manner, they are called finite ordinal numbers. The particular ordinal number which is assigned to the last term of a progression, if the progression has a last term, is called <u>the</u> ordinal number of that progression.* The first theory about the nature and meaning of the number concept, the theory we shall consider in the present chapter, says that natural numbers are basically ordinal numbers.

Now, suppose we have several collections of terms whose members are not arranged in any fixed order. An obvious property of such collections is the populousness or manyness of their respective terms. The numbers 1, 2, 3, . . . also may be used to represent this property of unordered collections. Whenever natural numbers are used in this manner, they are called cardinal numbers. The specific cardinal number assigned to the last term in an unordered collection, if the collection has a last term, is called <u>the</u> cardinal number of the collection. A second type of number theory, to be considered in Chapter 4, posits that natural numbers may be construed as cardinal numbers. A third type of theory, to be considered in Chapter 5, posits that natural numbers may be construed as both ordinal and cardinal numbers.

Whenever a collection of terms forms a progression, there is an important isomorphism between ordinal and cardinal numbers; that is, the ordinal number of any such collection automatically gives that cardinal number of the collection. Moreover, the ordinal number of any specific term in the collection automatically gives the cardinal number of the terms up to and including that point in the progression. These two facts led many nineteenth-century mathematicians (for example, Georg Cantor, Leopold Kronecker, and Richard Dedekind) to suppose that the ordinals are logically more fundamental than the cardinals. Ultimately, this belief provided the impetus for the doctrine that natural numbers may be identified with ordinal numbers.

The view of the number concept to be considered in this chapter maintains, generally speaking, that the numbers 1, 2, 3, . . . may

---

*The ordinal number 1 is the class of all classes containing a single term, the ordinal number 2 is the class of all classes containing an ordered pair of terms, and so on.

be reduced to values in the domains of the variable terms of a particular family of relations—relations of order. It is this general type of relation that always obtains between the terms of any collection which is also a progression. Historically, this approach to the number concept is by far the oldest of the three doctrines we shall be considering. There is evidence that the Greeks, when they thought at all about the underlying meaning of numbers, thought of them as relational entities. Certain sections of Euclid's Elements, written during the third century B. C., seem to presuppose a relational view of numbers.* Likewise, there is evidence that the Greek atomistic philosophers Leucippus (fifth century B. C.) and Democritus (fifth or fourth century B. C.) thought of numbers in this manner.

Relational theories of number originate in the observation that the most striking feature of the system of natural numbers is its inherent ordering. The terms of this system form a simple infinite progression and any finite subset of terms (for example, all even numbers less than 1,000) forms a finite progression. When we say, write, or read "1, 2, 3," and so on, they pass through our awareness in the same immutable order. Similarly, when we assign numbers to objects in the real world, the numbers always are assigned in the same fixed order. Intuitively, this suggests that the property of order—or, more precisely, the special ordering which characterizes the natural numbers—may be the key to unlocking the meaning of the number concept. This conjecture serves as the historical origin for doctrines in which the natural numbers are identified with the finite ordinal numbers.

Order, as one might suspect, is a purely relational idea. In the language of Chapter 2, it is concerned with a special family of propositional functions in two unknowns. To know what it means to define number in terms of order, it is obvious that we shall have to discover what type of relation order is. This is one of the chief tasks of any relational characterization of number.

NUMBER AS ORDER

The relational approach dates back at least to the Greeks and, consequently, is implicit in the work of many early mathematicians. During the nineteenth century, in particular, it is found in the work

---

*Referred to here are Definition 5 in Book V and the theory of proportions in Book VII. The author is indebted to Filmer S. C. Northrop for first drawing his attention to the fact that Euclid, albeit implicitly, takes a relational view of the natural numbers.

of Cantor, Helmholtz, Kronecker, and others. Today, however, the definition of numbers in terms of the relation of order usually is associated with the work of two mathematicians—Richard Dedekind and Guiseppe Peano. The relational approach first becomes explicit in their work. Of the two, Peano's contributions have been the more influential. This owes partly to the opaqueness of Dedekind's most important treatise on number, a fact which made it difficult for his contemporaries to determine precisely what it was that he was saying. It also owes partly to the favorable treatment that Russell gave to Peano's work in the Principles of Mathematics and Principia mathematica.

Before reviewing the main features of both Dedekind's and Peano's views, it would be helpful to offer an encapsulization of the general framework from which their work emerges. Both identified the natural numbers with the ordinals. This amounts to defining the number concept by appealing to the fact that natural numbers form a progression and that, as we have just seen, they may be used to represent the terms in any finite or simple infinite progression. Since the concept of progression obviously is fundamental to the relational viewpoint, we require a somewhat more precise definition of this notion than has heretofore been necessary. A progression or series is an ordered collection of terms in which each successive term in the series may be derived by applying a constant law of some sort to the immediately preceding term. The following, for example, are progressions in this sense: 0, 4, 8, 12, . . . ; 2, 4, 16, 256, ... The laws on which these progressions are based are "adding 4," "multiplying by 1/2," and "squaring," respectively. The particular law on which the series 1, 2, 3, . . . of natural numbers is based is "adding 1." More explicitly, the series of natural numbers may be described as that particular simple infinite progression which begins with 1, adds 1 to the first term to obtain the second term, adds 1 to the second term to obtain the third term, and so on.

The progressions of elementary mathematics are of two general types—"arithmetical progressions" and "geometrical progressions." An arithmetical progression is one in which the generative law is additive—that is, each term may be obtained by adding some fixed value to the immediately preceding term. The fixed value is called the common difference. All arithmetical progressions may be represented by the general progression

$$x,\ (x + a),\ (x + 2a),\ (x + 3a),\ \ldots \qquad (3.1)$$

where the constant a is the common difference. The individual terms in arithmetical progressions differ only in the coefficient of the constant a. The coefficient is 0 for the first term, 1 for the second term,

2 for the third term, and so on. It is easy to see that the coefficient for any arbitrary term in an arithmetical progression must be one less than the ordinal number of that term. A geometrical progression, in contrast, is one in which the generative law is multiplicative—each term may be obtained by multiplying the immediately preceding one by some fixed value. The fixed value is called the common ratio. Any geometrical progression may be represented by the general expression

$$x, xb, xb^2, xb^3, \ldots \qquad (3.2)$$

where b is the common ratio. Individual terms in a geometrical progression differ only in the power to which the common ratio is raised. The power is 0 for the first term, 1 for the second term 2 for the third term, and so on. In general, the exponent of the constant is one less than the ordinal number of the given term.

Of the illustrative progressions mentioned above, the first obviously is arithmetical and the latter two obviously are geometrical. The natural numbers also are an arithmetical progression. They constitute that particular arithmetical progression which may be represented as

$$1, [1 + 1], [1 + 2(1)], [1 + 3(1)], \ldots \qquad (3.3)$$

They are, in other words, that arithmetical progression which begins with 1, has 1 as the common difference, and has no last term.

The Dedekind Theory

Today, Richard Dedekind is chiefly remembered as the mathematician who devised the so-called Dedekind Cut, the first widely accepted method of defining the real number line. He also is remembered, along with Karl Weierstrass, for having eliminated the bothersome problem of infinitesimals (quantities which, though infinitely small, nevertheless are greater than zero) from mathematics by showing how to translate differential and integral calculus into the language of ordinary arithmetic without any loss of meaning. However, we are interested in Dedekind because he was the first modern mathematician to propose a completely relational theory of number. The theory appears in _Was sind und was sollen die zahlen_ (_The Nature and Meaning of Numbers_), a "memoir" first published in 1887. Although a translation of this pamphlet has been widely available since 1901, the details of the Dedekind theory are not well known among English-speaking scientists. This is largely attributable, it seems, to the obscure and difficult manner in which the pamphlet's central ideas are developed. Fortuunately, Russell provided a lucid and faithful reconstruction of the Dedekind theory in Principles of Mathematics. Those who desire a more complete discussion of Dedekind's views than the brief synopsis which follows here are advised to read Russell's account before taking up the original.

It happens that the whole of arithmetic and, therefore, all the rest of classical mathematics follows from the notion of simple infinite progression. Appearances to the contrary notwithstanding, arithmetic is not so much concerned with the system of natural numbers as it is with simple infinite progressions. Dedekind's most important achievement was to demonstrate the truth of this proposal. He showed that the first laws of arithmetic actually are concerned with the terms of simple infinite progressions and not with natural numbers per se. Although it is true that these laws and the theorems which follow from them, as usually stated, involve natural numbers, it is the fact that natural numbers comprise a simple infinite progression which makes these statements true. This fact, in turn, entails that the theorems of arithmetic are true for any progression—not merely for 1, 2, 3, ...

The key element in Dedekind's proof was a demonstration that the first laws of arithmetic depend directly on a certain rule of inference which happens to be valid for all simple infinite progressions—that is, "mathematical induction." According to this rule, if there is any collection of terms to which belong (1) the first term of some progression and (2) the term immediately following any term of the progression, then every term of the progression belongs to the collection. The meaning of this somewhat difficult rule is made clearer by translating it into the language of propositional functions. Suppose there is some propositional function, R_a. As is always the case with such a function, the truth of R_a will vary depending on the specific values assigned to a. It will be true for some values, false for some, and meaningless for others. Suppose that R_a is assigned a term from a simple infinite progression as a value and suppose a' denotes the term which immediately follows the assigned value, whatever the assigned value may be. Suppose it is possible to show, first, that R_a is true if a is the first term of the progression and, second, that $R_{a'}$ is true whenever R_a is true. If these two premises can be established, then, by the principle of mathematical induction, we may conclude that R_a is true for every term in the progression. Thus, an important general fact about every simple infinite progression is that any statement which is true of the first term and true of the immediate successor of each term is also true of every term in the progression. It is precisely this general property of progressions which, as Dedekind was able to show, makes the theorems of arithmetic true. Later, when we consider Peano's views, we shall see how this property may be used to axiomatize the system of natural numbers.

Thus, it is only the fact that natural numbers form a progression that appears to be relevant to ordinary arithmetic. Since ordinal numbers are what the terms of all progressions have in common, Dedekind was led to identify natural numbers with ordinal numbers: "These elements are called natural numbers or ordinal numbers or simply number" (Dedekind 1887: 68). The fact that arithmetic depends solely on the ordinal properties of the natural numbers also led Dedekind to conclude that ordinal numbers are logically more fundamental than cardinal numbers. On this particular point, we already have seen that the ordinal number associated with any given term in a progression gives the cardinal number of the collection of terms up to and including the given term. For our purposes, this is the most important fact about the Dedekind theory. It suggests that, whatever else the natural numbers may be, they are, first of all, a progression.

This brings us to the problem of characterizing the logical type of relation, R_{ab}, which always obtains between the terms of any simple infinite progression. When we say, as in Dedekind's theory, that the system of natural numbers springs from the idea of simple infinite progression, we do not have any particular progression in mind. We are concerned, rather, with translating the generic idea of progression into the logical language of Chapter 2. This means that we are not concerned with any particular ordered collection of terms or any particular generative law. What is crucial for arithmetic is that we are able to say what generic type of relation R_{ab} must always be—that is, what, if any, are the common relational properties of laws such as "greater than by 1," "less than by 2," and "square of." What general types of relations invariably arrange the collections of terms which satisfy them into progressions? This is the particular family of relations on which arithmetic depends.

The family of relations whose members always arrange the values which satisfy them into progressions are "transitive-asymmetrical" relations. The values in the domains of the variable terms of any such relation always compose a progression. Of the three illustrative progressions mentioned earlier, each generative law implies, among other things, a transitive-asymmetrical relation. All three generative laws ("adding 4," "multiplying by 1/2," "squaring"), for example, entail the relation "successor of." This relation is transitive because, for any three terms, a, b, and c, from any of the three collections, if "b is the successor of a" and "c is the successor of b" both are true, then "c is the successor of a" also must be true. The relation also is asymmetrical because, for any two terms a and b from any of the three domains, if "b is the successor of a" is true then "a is the successor of b" cannot be true, and conversely. Whenever the domain of the transitive-asymmetrical relation implied by some particular generative law has a first term and no last term, the relation arranges the terms into a simple infinite progression. Whenever the domain of such a relation has a last term, the relation arranges the terms in to a finite progression. The aforementioned progressions, which all involve natural numbers, are cases in point.

The important fact to bear in mind about the Dedekind theory is that, by identifying the natural numbers with the terms of finite and infinite progressions, the concept of natural number is reduced to the more fundamental concept of transitive-asymmetrical relation. The latter concept is more fundamental for two reasons. First, we saw in Chapter 2 that it belongs to logic, while finite and infinite progressions belong to mathematics. Second, although the generative law of any finite or infinite progression automatically gives a transitive-asymmetrical relation, the converse is not true. That is, there are transitive-asymmetrical relations to which there correspond no mathematical progressions. For example, consider the relation "order of introduction" and, as values of its variable terms, consider the collection of characters in any of Balzac's novel. The relation in question obviously is both transitive and asymmetrical. It arranges the characters in the order of their initial appearance. But the ordering which results obviously is not a progression, since mathematics is not concerned with characters in Balzac's novels.

The Peano Theory

Although Peano's and Dedekind's theories do not differ in principle, it has been pointed out that Peano's theory is considerably better known. In particular, it has been favored by the members of a school of mathematical foundations known as "formalism." Twentieth-century thinking about the foundations of mathematics has been dominated by three great schools. We already have encountered two of them, logicism and intuitionism, in Chapter 2. As we saw, logicists believe that mathematics is simply logic and have tried, therefore, to say precisely what logic is. Intuitionists, like Pythagoras, believe that mathematics springs from the concept of natural number. Formalism, founded by the German mathematician David Hilbert, advances yet another hypothesis. According to formalism, mathematics should be thought of as a game played on sheets of paper with certain meaningless symbols. A rather more precise synopsis of this doctrine was offered in 1931 by Johann von Neumann:

> Even if the statements of classical mathematics should turn out to be false as to content, nevertheless, classical mathematics involves an internally closed procedure which operates according to fixed rules known to all mathemati-

> cians and which consists basically in constructing successively certain combinations of primitive symbols which are considered 'correct' or 'proved.'. . . If we wish to prove the validity of classical mathematics . . . then we should investigate, not statements, but methods of proof. We must regard classical mathematics as a combinatorial game played with the primitive symbols, and we must determine in a finitary combinatorial way to which combinations of primitive symbols the construction methods or 'proofs' lead (cited in Benacerraf and Putnam 1964: 50-51).

As games go, mathematics has proved to be somewhat complicated and challenging. For this reason, formalists favor comparing it with chess rather than old maid. What is important about any game, no matter how complicated, is the rules by which it is played. So it is with mathematics. As mathematicians, we should not be particularly interested in plumbing the meaning of the entities in mathematical formulas. Ultimately, they are nothing more than meaningless symbols. What we should be concerned with, instead, is the rules by which the game of mathematics is played. Applying this argument to the specific problem of formulating a theory of number, it is not the meaning of individual natural numbers that is in doubt. What is in doubt is the rule or rules which govern the system. This brings us back to the Peano theory. From the formalist perspective, this theory accomplishes essentially two things. First, it tells us that the pieces on the chessboard, the most primitive of all the primitive symbols with which mathematicians play, are "1," "2," "3,". . . . There is nothing special or inevitable about these symbols. We could just as easily use others. Second, the theory tells us that the most fundamental rule of the game of mathematics—the only rule, in fact, that we absolutely cannot get along without—is that we must keep our symbols arranged in a fixed order. If we do, the theorems of arithmetic and those of the rest of classical mathematics follow. Otherwise, they do not. Since it happens that we have provided ourselves with infinitely many primitive symbols, it turns out that, as long as we obey the order rule, they comprise what we have been calling a simple infinite progression. It is obvious that this general approach tends to denude individual natural numbers of any intrinsic concrete meaning. From a formalist perspective, this is a very important virtue. Below, we shall see that logicists find this an equally important shortcoming.

Peano's theory consists of three undefined concepts and five axioms which make use of the three concepts. For the sake of convenience, the theory may be regarded either as an axiomatization of the notion of simple infinite progression or, alternatively, as an axiomatization of the counting operation by which such progressions are generated. This axiomatization also may be viewed as leading to the conclusion that any countably infinite (simple infinite) progression implies a transitive-asymmetrical relation. Peano's undefined concepts are "zero," "finite integers," and "successor of." By the latter concept, Peano actually understood "immediate" successor of. By the first concept, Peano understood the first term in the series of natural numbers. To maintain continuity with previous chapters, in which the natural numbers have been characterized as the series 1, 2, 3, . . . rather than the series 0, 1, 2 . . . , the undefined concept "one" will be substituted for Peano's zero. Finally, also for the sake of continuity, the undefined concept "natural number" will be substituted for Peano's finite integer. Peano's axioms, using the revised undefined concepts, may now be stated:

1. One is a natural number.
2. If x is any natural number, then the immediate successor of x also is a natural number.
3. If x and y are any two natural numbers and if x and y both have the same natural number as their immediate successor, then x and y are the same natural number. That is, two different natural numbers may not have the same immediate successor.
4. One is not the immediate successor of any natural number.
5. If there is some collection to which one belongs and to which the immediate successor of every natural number belongs, then all natural numbers belong to this collection.

The Peano theory first appeared in the 1899 edition of Formulaire de mathématiques. Two essential facts about the theory also were proved. First, it was shown that all of arithmetic follows from the axioms; assuming that there are interpretations of the three undefined concepts which make all five axioms true, the first laws of arithmetic and the theorems which follow from them also are true. When "one" means "1," "natural number" means "1, 2, 3, . . . , " and "immediate successor of" means "greater than by 1," all five axioms and, consequently, the rest of arithmetic are true. Second, and equally important, Peano and one of his collaborators (Padoa) showed that the five axioms in question are the absolute minimum required to make arithmetic possible. That is, none of the five axioms is implied by one or more of the remaining axioms. Hence, each axiom may be viewed as completely independent of the other four. Peano and Padoa demonstrated this by showing that, for every possible group of four axioms, there is some interpretation of the undefined concepts which makes the four axioms true but makes the remaining axiom false. For example, suppose we define "one" and "natural number" as above, but we take "immediate successor of" to mean "greater than by 2." Under this particular interpretation of the undefined concepts, the first four axioms are true. But the fifth axiom is false because the series 2, 4, 6, . . . , all of whose members belong to the series of natural numbers, is left out of account. As a further illustration, suppose we define "one" and "immediate successor of" as above, but we take "natural number" to mean "2, 3, 4, ..." Now, the last four axioms are true but the first one is not.

Peano recognized that there are interpretations of the undefined concepts other than the usual ones ("1," "1, 2, 3, . . . ," and "greater than by 1") which make all five axioms true. In fact, any collection of terms that (1) has a first term, (2) has no last term, (3) has no repeated terms, and (4) is such that any term can be reached from the first term in a finite number of steps makes all the axioms true. The collections of terms which have just these properties are what we have been calling simple infinite progressions. Peano argued that the concept of number—in particular, the system of natural numbers—may be thought of as the property which is shared by all collections which satisfy the preceding four conditions: "Number is what is obtained from all these systems by abstraction; in other words, number is the system which has all the properties enunciated in the primitive propositions, and those only" (Quoted from Russell's, 1903, p. 125, translation). We already have seen that the one property shared by all simple infinite progressions—and, therefore, by all systems which satisfy Peano's five axioms—is that they imply a transitive-asymmetrical relation of some sort. For our purposes, therefore, the general result of Peano's theory is the same as Dedekind's. The concept of number is reduced to a special case of the family of propositional functions in two unknowns which are both transitive and asymmetrical. Any interpretation of Peano's three undefined concepts that makes the five axioms true also entails a propositional function of the form R_{xy} which is easily shown to be transitive-asymmetrical. To illustrate, suppose that "one," "natural number," and "immediate successor of" are assigned their usual interpretations. Among other things, these interpretations imply a propositional function of the form "x follows y." The first variable term takes as values terms from the series 2, 3, 4, . . . , while the second variable term takes as values terms from the series 1, 2, 3, ... The third axiom insures that R_{xy} is both transitive and asymmetrical. If two different natural numbers never have the same immediate successor, then, given any three natural numbers, a, b, and c, R_{ac} must be true whenever R_{ab} and R_{bc} both are true. Note that R_{ab} and R_{bc} can both be true if and only if $a \neq b \neq c$. If two different natural numbers never have the same immediate successor, then, given any two natural numbers, a and b, R_{ba} cannot be true whenever R_{ab} is true. Again, either premise can be true if and only if $a \neq b$.

DISCUSSION

The general doctrine which we have been considering up to this point is fundamentally "holistic" rather than "atomistic." That is, it emphasizes a generic property which is true of natural numbers as a whole, and it does not give more than passing attention to the meaning of individual numbers. The general effect of both the Dedekind and Peano theories is to reduce natural numbers to values in the domains of certain types of relations (transitive-asymmetrical relations). As we saw with Peano's theory, there are values other than natural numbers which satisfy transitive-asymmetrical relations. Individual natural numbers, therefore, take such meaning as they may have from the relations into which they enter. Numbers, in their most basic sense, are not "things." They refer to a certain family of relations. It is in just this sense that the relational approach to numbers is holistic. Concerning that particular subgroup of transitive-asymmetrical relations from which simple infinite progressions result, all of the remaining terms in any such progression are given once we know the particular generating relation and have fixed an arbitrary starting point. In the case of the particular simple infinite progression 1, 2, 3, . . . , the relation happens to be "greater than by one" and the arbitrary starting point happens to be "one." The mere fact that the terms which result from applying this particular relation repeatedly to this particular starting point are universally known as natural numbers is simply an accident of semantics and should not be viewed as having any deeper mathematical significance.

Historically, two objections have been lodged against the preceding characterization of the natural numbers. Both originated, or at least were popularized, by Bertrand Russell, and they are regarded by many as sufficient to necessitate a quite different approach to defining the number concept. The first objection is broadly concerned with the concrete specificity of the relational definition. Its point of departure is the aforementioned fact that values other than natural numbers can satisfy the definition. The second objection is concerned with the fact that an arbitrary starting point is used in the relational construction of the series of natural numbers. It is argued that this approach leaves the number "1" in 1, 2, 3, . . . undefined. Since the remaining numbers in the series are defined in terms of 1, it is argued that these numbers, too, are undefined.

Although the first argument applies to both Dedekind's and Peano's theories, Russell (1903, 1919) originally devised it as a reply to the Peano theory. It has already been noted that the argument deals with the fact that numbers other than 1, 2, 3, . . . (and, indeed, things other than numbers) satisfy the relational definition. Explicitly, Russell exhibited several interpretations of Peano's undefined concepts other than the usual ones which make all five axioms true. From this it follows that the axioms are true for systems other than the natural numbers and, hence, the theory is not specific to our everyday meaning of the natural numbers:

> Peano's three primitive ideas . . . are capable of an infinite number of different interpretations, all of which will satisfy the five primitive propositions. . . . (1) Let '0' be taken to mean 100, and let 'number' be taken to mean the numbers from 100 onward in the series of natural numbers. Then all our primitive propositions are satisfied, even the fourth, for, though 100 is the successor of 99, 99 is not a 'number' in the sense which we are now giving to the word 'number'. . . . (2) Let '0' have its usual meaning, but let 'number' mean what we usually call 'even numbers', and let the 'successor' of a number be what results from adding two to it. Then '1' will stand for the number two, '2' will stand for the number four, and so on. . . . (3) Let '0' mean the number one, let 'number' mean the set
>
> 1, 1/2, 1/4, 1/8, 1/16, . . .
>
> and let 'successor' mean 'half'. Then all Peano's five axioms will be true of this set. . . . In fact, given any series
>
> $x_0, x_1, x_2, x_3, \ldots x_n, \ldots$
>
> which is endless, contains no repetitions, has a beginning, and has no terms that cannot be reached from the beginning in a finite number of steps, we have a set of terms verifying Peano's axioms (Russell 1919: 7-8).

We already have seen that Peano explicitly acknowledged that progressions other than the natural numbers—in fact, any of the infinite number of simple infinite progressions—will satisfy the five axioms. However, Russell argues that this entails a problem which Peano did not recognize:

> Every [simple infinite] progression verifies Peano's five axioms. . . . Any progression may be taken as the basis of pure mathematics: we may give the name '0' to its first term, the name 'number' to the whole set of its terms, and the name 'successor' to the next in the progression. The progression need not be composed of numbers. . . . Each different progression will give rise to a different interpre-

> tation of all the propositions of traditional pure mathematics. . . . In Peano's system there is nothing to enable us to distinguish between these different interpretations of his primitive ideas. It is assumed that we know what is meant by '0,' and that we shall not suppose that this symbol means 100 or Cleopatra's Needle. . . . This point, that '0' and 'number' and 'successor' cannot be defined by means of Peano's five axioms, but must be independently understood, is important. We want our numbers not merely to verify mathematical formulae, but to apply in the right way to common objects. We want to have ten fingers and two eyes and one nose. A system in which '1' meant 100, and '2' meant 101, and so on, might be all right for pure mathematics, but would not suit daily life. We want '0' and 'number' and 'successor' to have meanings which will give us the right allowance of fingers and eyes and noses. . . . We cannot secure that this shall be the case with Peano's method. . . . (Russell 1919: 8-9; emphasis added).

The sense of the first objection, then, is this. The reduction of the system of natural numbers to the family of transitive-asymmetrical relations suffices to make the theorems of arithmetic true. However, we desire something more from a definition than such an abstract payoff. We require that each term in the series 1, 2, 3, . . . have a single concrete meaning. Here. "single," also means "unique." More explicitly, we should require that the natural numbers have the definite meaning which corresponds to the manner in which we use them when we enumerate objects in the real world. In the relational approach, natural numbers have no such concrete meaning. They are, as we have seen, simply "variables" in the general sense we used to define this notion in Chapter 2. If we adopt the relational approach, we shall not have shown that the statements of arithmetic and, more important, their counterparts in the real world are true for a special collection of terms which we have agreed to call natural numbers. Instead, we shall have demonstrated that these statements are true for any and all collections which formally imply a transitive-asymmetrical relation. We will be left in the dark as to whether or not there are any such collections in the real world.

The second objection is concerned with a particular member of the series of natural numbers, that is, 1. The relational method of defining natural numbers begins with an undefined unit term of some sort. It then obtains 2 from 1 by adding the unit term to itself once. It obtains 3 from 1 by adding the unit term to itself twice. In general, it obtains n from 1 by adding the unit term to itself n - 1 times. Actually, of course, the relational definition says that 1 equals the first term in any collection that has the formal properties which we have been discussing. Hence, 1 cannot be just any "unit." However, we have seen that, insofar as the relational definition is concerned, 1 can mean many things other than its usual common sense meaning (for example, 100 or 100,000). When 1 does mean a unit other than its common sense meaning, then the meanings of the other natural numbers are similarly altered.

It is easy to see that this objection is simply a special case of the initial criticism. What is being said is that we should require some definite—that is, unique—starting point for the series of natural numbers. If we can define 1 in such a way that it refers to our common sense meaning of the term, then the remaining natural numbers also will have their everyday meanings. If, instead, we allow 1 to mean the first term of any simple infinite progression, then there is no guarantee that 2, 3, 4, . . . will have their everyday meanings.

It does not seem especially difficult to construct counterarguments for the preceding objections.* Both of the above objections to the relational approach say (or presuppose) the same thing: We ought to have a definition which assigns to the series 1, 2, 3, . . . a constant meaning rather than a variable one. In addition to assigning a constant meaning, the particular constant meaning should be the same one that we use in everyday life when we enumerate objects. Thus, every term in the series is to have a nonvariable meaning and that meaning is to be the same as the term's meaning in everyday usage. Since each term in the series is defined by adding 1 to itself repeatedly, the way to insure that each term has its everyday meaning is to assign 1 its everyday meaning. In the relational approach, however, 1 can mean the first term of any progression. Consequently, the subsequent terms of the series also have variable meanings.

---

*Before proceeding to rebut the objections, however, the author should note where his own biases lie. As will become clearer subsequently, he believes that the relational approach to the number concept is by far the most satisfactory of the three approaches that will be considered and that there are sound logical and psychological reasons for this preference; it is intended to make them clearer as we proceed. For the present, however, honesty dictates that he note his prorelational inclination. Quite obviously, this inclination leads the author to portray the relational viewpoint favorably and, where possible, to construct counterarguments for objections such as the two we have just considered. While he cannot believe that his biases affect the validity of such counterarguments, they undoubtedly have caused him to select what seem to be the more telling ones.

Against Russell's general proposal that a theory of number, to be deemed acceptable, must assign 1, 2, 3, . . . their meaning in everyday usage, at least two counterproposals are possible. First, and most importantly, we must take care to distinguish between criticisms of the relational theory which are legitimate mathematical or logical objections and criticisms which reflect philosophical or psychological doctrines. The proposal that natural numbers should be assigned their everyday meanings definitely is a criticism of the latter sort. It is not suggested, as part of this criticism, that defining natural numbers as values of the variable terms of transitive-asymmetrical relations presents any mathematical problems. Quite to the contrary, it is acknowledged that such a definition suffices to make the first laws of arithmetic true. Therefore, it is very difficult to see where this criticism leads us and how it relates to our original aims. Recalling the discussion of number-theory goals in Chapter 2, it will be remembered that the principal motivation for seeking a definition of the number concept in terms of more primitive notions is purely mathematical. We saw that defining the number concept formally first became important as a means of making mathematics itself possible. The theorems of classical mathematics depend on the theorems of arithmetic, and the theorems of arithmetic, as usually stated, depend on the natural numbers. Hence, we require a definition of the natural numbers from which all the theorems of arithmetic follow. The relational definition satisfies this requirement. Without the existence of arithmetic and the rest of classical mathematics, a formal definition of the number concept would not be any more necessary than a formal definition of purely psychological concepts such as love, justice, and sympathy.

The important point to bear in mind here is that our need for a definition of number does not ultimately derive from the fact that numbers, like football and poetry, are psychologically important or the objects of philosophical speculation. It derives, rather, from the central position of numbers in classical mathematics. Therefore, the one absolutely essential criterion for any number theory is that it makes mathematics possible. No matter how ardently we may desire a theory which squares with our common-sense conception of natural number, we must not permit this fact to obscure the supremacy of the mathematical criterion. Now, we already know that the theorems of arithmetic depend on the fact that the numerical symbols which appear in them are always arranged in a fixed order. Any concrete "meaning" that these symbols may have is strictly beside the point. Since our original interest in numbers stems not from whatever status they may have as independent concrete entities but from their role in mathematics, it is argued here that the notion of "ordering" or "transitive-asymmetrical relation" is a perfectly admissible characterization of "number." To maintain otherwise would seem to imply either a vaguely Pythagorean doctrine in which natural numbers are possessed of a metaphysical existence or a vaguely utilitarian conception in which natural numbers are nothing more than generalizations from everyday experience. Interestingly, Bertrand Russell subscribed to the former doctrine at the time he formulated the two objections considered above. Many years later, he confessed that "at the time when I wrote the 'Principles,' I shared with Frege a belief in the Platonic reality of numbers, which in my imagination, peopled the timeless realm of Being. It was a comforting faith, which I later abandoned with regret" (1937: x). In any case, regardless of whether one's predispositions are Pythagorean or utilitarian when it comes to numbers, the point is precisely the same. Any criticism of the relational theory which is predicated on either view is purely philosophical. Moreover, such objections tend to obscure the fact that the only essential criterion which must be met is that of making mathematics possible.

This brings us to a second objection to the proposal: than an acceptable number theory must invariably assign 1, 2, 3, . . . their everyday meanings. Suppose, for the sake of argument, we stipulate that this objection is correct. That is, our aesthetic preference for a theory which embodies our common-sense conception of the natural numbers is just that, a preference. Having admitted this much, suppose that we nevertheless proceed to act on our preference and we construct a definition which satisfies it—much as Russell did when he devised the theory discussed in the next chapter. We still have to satisfy the criterion of a definition which makes mathematics possible. What guarantee do we have that a definition which incorporates only the everyday usage of 1, 2, 3, . . . will satisfy this criterion? We have none. There is absolutely no a priori reason for supposing that such a definition will be mathematically satisfactory. The fact that, as we have seen, the theorems of arithmetic do not depend on the natural numbers per se raises further a fortiori doubts. It is even possible that, instead of making mathematics possible, such a definition might introduce insuperable difficulties.

What has just been said could easily be misinterpreted as a neo-Kantian or, more generally, idealistic stance. Explicitly, these remarks might be taken to imply a doctrine in which mathematics somehow results from something other than everyday experience. Since scientists would find such a doctrine both abhorrent and silly, this misinterpretation must be quickly corrected. It is not suggested that, for example, the statements of arithmetic do not originate in statements about the real world. Although mathematics requires that we behave as though we knew certain things independent of our experience, the basic theorems of arithmetic—most certainly the first laws—are inductive generalizations from everyday experience. Indeed, the whole of Chapter 1 was concerned to show, among other things, that arithmetic, algebra, and the number concept are products of a gradual evolutionary process in which, despite Pythagorean protests to the contrary, the principal influences have been sociocultural. Thus, in so far as mathematics is concerned, idealism is to be rejected in favor of a view which is historical and utilitarian. However, from the general fact that mathematics springs from common sense, it certainly does not follow that it springs, in particular, from our everyday understanding of the numbers 1, 2, 3, ... It is equally reasonable to suppose that mathematics is rooted in some other common-sense idea. What the relational approach to number ultimately seems to tell us is that it is rooted in the experience of ordering concrete objects according to common transitive-asymmetrical relations ("longer than," "heavier than," "taller than," "darker than," and so on).

History also teaches us that our common-sense conception of 1, 2, 3, . . . should not be the final arbiter of number theories. Recall here our discussion in Chapter 1, especially our discussion of the ancient period. Recall, in particular, that for arithmetic and algebra to evolve into reliable computational sciences it was necessary to adopt an abstract attitude toward numbers. We saw that this was the principal difference between the Babylonian and Egyptian approaches. Consequently, the Babylonians were able to construct precise computational methods while the Egyptians never progressed beyond the level of rough-and-ready estimation. Further, the subsequent struggle to broaden the number concept to include such things as negative numbers and complex numbers was largely a struggle to free the number concept from the tyranny of the everyday meaning of 1, 2, 3, ... In view of the critical role played by the abstract attitude toward numbers in the evolution of mathematics, there seems to be no reason to suppose that the reverse of this attitude should be adopted as an ultimate criterion for judging number theories. Once again, this may seem a neo-Kantian or idealistic stance. Nothing could be further from the truth. It is not suggested that the meaning of numbers is, in some mysterious way, given a priori. What is suggested is that there is a historical dichotomy between the concrete and abstract views of number which merits consideration. Historically, it seems axiomatic that the number concept springs from everyday experience. However, beyond a certain uncouth level in the development of arithmetic and algebra, the common-sense interpretation of number becomes distinctly counterproductive. In fact, if mathematics is to progress beyond this level, we must completely suppress this interpretation. This suggests that, while the proposal that a number theory should assign 1, 2, 3, . . . their everyday meanings may be appropriate for the earlier phases in the development of mathematics, it is not necessarily appropriate today. If (1) our first goal is to make mathematics possible and (2) in order to do this, it was historically necessary to adopt an abstract attitude toward numbers, why should we now, milleniums later, suddenly wish to return to a concrete attitude? This seems a curiously reactionary desire.

4

THE CLASSIFICATORY VIEW OF NUMBERS

> Returning now to the definition of number, it is clear that number is a way of bringing together certain collections, namely, those that have a given number of terms.
>
> Bertrand Russell

We turn now to the theory-of-classes counterpart of the approach to number considered in the preceding chapter. Whereas the relational approach defines the natural numbers as some of the values of the variable terms of a certain family of relations, the classificatory approach defines them as the only values of a particular family of classes. And whereas the relational theory begins by identifying the natural numbers with the ordinals, the theory to be considered below begins by identifying the natural numbers with the cardinals. Hence, number is associated with the manyness of collections of terms rather than the positions of progressions of terms. Each natural number is specifically defined as the class of all those classes which possess that number of terms.

The theory of "number as class" was independently discovered by Gottlob Frege, sometime before 1884, and by Bertrand Russell during the spring of 1901. We saw in Chapter 2 that this theory usually is associated with the logicist school founded by Russell and Alfred North Whitehead. Historically, however, Frege's version of the theory has priority over Russell's. This version appeared in Frege's monograph Die Grundlagen der Arithmetik (The Foundations of Arithmetic) (1884). Although historians of mathematics place considerable emphasis on first publication of any new theory (for example, Bell 1940), our present awareness of this particular theory is due almost entirely to Russell's independent discovery and subsequent expositions of it in the Principles of Mathematics (1903), Principia mathematica (1910-1913), and Introduction to Mathematical Philosophy (1919). The latter work, written during Russell's six-month imprisonment for pacifism in 1918, is primarily responsible for making the theory widely known outside of technical mathematical circles. The first chapter of this book, in which the theory appears, is easily understood by readers with no mathematical training beyond algebra. * The problem with Frege's version of the theory is that he, like Dedekind, shared the nineteenth-century German talent for obscurantism. Frege's writing is far more opaque than Dedekind's and some of the denser passages of *Die Grundlagen* rival the best efforts of idealist philosophers. Russell once admitted that although he had read *Die Grundlagen* prior to writing the *Principles*, he did not grasp what Frege was saying until he managed to discover the same theory independently. If Russell, at the height of his powers, could not decode Frege, one can well imagine the difficulties encountered by less gifted mathematicians and logicians of his era. In any case, it was only after Russell had formulated the theory in comprehensible language that Frege's priority became known. In fact, without Russell's *Principles*, where Frege's priority is scrupulously documented in an appendix prepared after the rest of the book had gone to press, the fact that the theory first appeared in *Die Grundlagen* might have remained unknown.

Given these facts, and for the sake of brevity, we shall consider only Russell's version of the theory that numbers are classes rather than relations. It is true that there are some differences between Russell's and Frege's respective accounts of the theory. These differences stem primarily from the fact that Russell and Frege employed slightly different definitions of the underlying concept of class (Russell 1903: 511-18). Frege preferred to define classes by referring to their intensions. Russell, however, adopted the more formal definition discussed in Chapter 2, in which classes are concerned with propositional functions in one unknown. The latter definition has the effect of making classes synonymous with their extensions. Frege also employed a somewhat vaguer definition of the membership relation, $\in$, than its formal definition as a certain type of propositional function in two unknowns. For our part, these are very technical differences which may be safely ignored. Both versions of the theory agree on the three important points. First, the idea of number springs from the idea of

---

*Interestingly, there does not seem to be a corresponding work which has made the relational theory widely known among general readers. Consequently, while modern adherents of the classificatory theory include groups other than mathematicians (philosophers, educators, scientists), the circle of adherents of the relational theory is confined almost entirely to people who are mathematicians by training.

manyness or populousness rather than from the idea of order. Second, the natural numbers are to be identified with the cardinals. Third, each natural number should be regarded as a certain kind of class. Whether such classes are to be defined by appealing to their intensions (Frege) or their extensions (Russell) is, for us, largely a matter of indifference.

THE THEORY

The specific version of Russell's theory which we shall consider is the first one, presented in Part II of the Principles. This particular treatment of the theory does not differ in any essential respect from subsequent treatments in Principia mathematica and Introduction to Mathematical Philosophy. In each case, there is a language difference but not a content difference. The general proposal which serves as the point of departure for the theory will be recalled from the discussion at the end of the preceding chapter; that is, an acceptable theory of number should assign the natural numbers one constant meaning and, moreover, this meaning should be the same as the one we use in everyday life. We have seen earlier that the logical grounds for this proposal are something less than compelling. In fact, they are somewhat dubious. For the present, however, let us suppose that we accept the proposal as correct.

Russell began his development of the theory with a criticism of the general concept of definition presupposed by the relational theory of number. According to this conception of definition, he argued, any concept is regarded as being satisfactorily defined whenever it has been shown to be a special case of certain constituent ideas. On the ground that wholes are not, as a general rule, precisely equal to the sum of their parts, Russell advocated a different approach to definition: "Given any set of notions, a term is definable by means of these notions when, and only when, it is the only term having to certain of these notions a certain relation which itself is one of the relations" (1903: 111). The first approach to definition is called "philosophical" and Russell's alternative approach is called "mathematical." Needless to say, these are not the usual meanings of these terms. The method of constructing the series of natural numbers discussed in Chapter 3, in which natural numbers are defined as that special arithmetical progression which begins with 1 and generates each number by adding 1 to its immediate predecessor, was given as an explicit illustration of philosophical definition. The definition was said to be unacceptable because, among other things, it produces a "tiresome difference" between 1 and other members of the series. By adopting the mathematical approach to definition, however, Russell claimed that it is possible to define 1 in the same manner as the other numbers.

To eliminate the difference between 1 and the other numbers, it was suggested that we should regard the property of number, generally, as a property of classes and that, in particular, we should regard the series of natural numbers as synonymous with the cardinals: "Numbers are, it will be admitted, applicable essentially to classes. . . . Thus when any class-concept [intension] is given, there is a certain number of individuals to which this class-concept is applicable, and the number may therefore be regarded as a property of the class" (1903: 112-13). Natural numbers, therefore, refer to how many terms a class contains, and the number of a class corresponds to a certain manyness. Thus, to define the natural numbers we must define the concept of cardinal number. The first step in the definition is to pose the question, "What does it mean to say that two classes have the same number?" in place of our earlier question, "What property of the natural numbers makes the theorems of arithmetic true ?" Russell answered the former question by appealing to the correspondence relation discussed in Chapter 2: "Two classes have the same number when their terms can be correlated one to one, so that any one term of either corresponds to one and only one term of the other" (1903: 113). In other words, given two classes, R_x and R_y, we may say that they have the "same number" (that is, "same manyness"), if, to every value x which satisfies R_x, there corresponds one and only one value of y which satisfies R_y, and conversely. This statement holds if, for example, R_x = "x is a monogamously wedded man" and R_y = "y is a polygamously wedded woman." Whenever the preceding statement is true for any two classes, R_x and R_y, the classes are said to be "similar." In other words, similarity is defined as a relation of the form R_{xy}, for which there is a one-to-one correspondence between the respective values of the two unknown terms which satisfy the function. The function "x is monogamously wedded to y" is an example of a similar relation. Note that this relation and all similar relations are transitive and symmetrical. Whenever the preceding statement is not true for any two classes, R_x and R_y, the classes are said to be "dissimilar." Dissimilarity, therefore, is a relation of the form R_{xy}, for which there is either a one-to-many or many-to-one or many-to-many correspondence between the respective values of the two unknown terms which satisfy the function. The function "x is polyandrously wedded to y" is an example of a dissimilar relation. The concept of dissimilarity was Russells' answer to the question, "What does it mean to say that two classes have different numbers?"

So far, the concept of "same number" ("same manyness") has been reduced to the concept of similarity which, in turn, is defined in terms of the relation of one-to-one correspondence. Also, the concept of "different number" ("different manyness") has been reduced to the concept of dissimilarity which, in turn, is defined in terms of the absence of the relation of one-to-one correspondence. Certain logical problems necessitate further refinements in this definition.

In the theory of classes, elements and classes are thought of as generating a natural hierarchy. At the first level, there are elements which are left undefined; elements belong to classes, which compose the second level in the hierarchy; classes belong to classes of classes, which compose the third level of the hierarchy; classes of classes belong to classes of classes of classes, which compose the fourth level of the hierarchy, and so on. It is easy to see that this hierarchy is a simple infinite progression: There is an arbitrary first term (elements), no repeated terms, no last term, and the nth term can always be reached from the first term in exactly n - "one" steps. Suppose we begin with any undefined element—a major league baseball player. Every player is a member of a team, which is a class. Every team belongs to a division, which is a class of classes. Every division belongs to a league, which is a class of classes of classes. Every league belongs to a corporation, which is a class of classes of classes of classes, and so on. Now, let us return to the definition of number. Up to this point, number has been viewed as a "property of the second level" of any hierarchy, which begins with elements proceeds to classes, then to classes of classes, and so forth. In the case of our illustration, "9" is regarded as a common property of the class of major league baseball teams.

It turns out that there are insuperable logical difficulties associated with thinking of numbers as common properties of classes. In particular, Russell observed that such a definition "does not show that only one object satisfies the definition . . . instead of obtaining *one* common property of similar classes, which is *the* number of the classes in question, we obtain a *class* of such properties" (1903: 114). The meaning of this statement is quite simple. It will be recalled that Russell set out to construct a definition which would assign number a single constant meaning. Defining numbers as properties of classes is unsatisfactory in this respect because, it is to be noted, classes have properties other than the number of their terms. Russell considered two methods of resolving this problem. The first, which he rejected, consisted of "defining as *the* number of a class the whole class of entities . . . to which all classes similar to the given class (and no other) have some many-one relation or other" (1903: 115). This solution was rejected on the quite proper ground that "all entities, without exception, belong to every such class, so that every class will have as its number the class of all entities of every sort and description" (ibid.). In other words, instead of having defined a particular cardinal number, one has defined the entire universe of elements. The second method, which Russell accepted, consisted of defining "as the number of a class the class of all classes similar to the given class. Membership of this class of classes . . . is a common property of all the similar classes and of no other; moreover every class of the set of similar classes has to the set a relation which it has to nothing else, and which every class has to its own set" (ibid.).

The implication of the second method is this: Each cardinal is to be regarded as a particular class of classes, that is, the class of all those classes between whose elements there is a one-to-one correspondence. A number, then, is a class of similar classes. It is sometimes supposed, especially by psychologists who study the role of numerical concepts in human reasoning (Piaget 1952) that the second method describes number as a property of classes of classes. That is, the property of number is simply removed from the second level of the aforementioned hierarchy and relocated at the third level. Although this is perhaps an understandable misinterpretation of the remarks quoted above, it is a misinterpretation nevertheless. Defining number as a property of classes of classes suffers from the same general defect as defining them as properties of classes, that is, classes of classes have properties other than the manyness of their constituent classes. Russell did not suggest that numbers be viewed as properties of classes of classes. He suggested, instead, that each number is a particular class of classes. Explicitly, the number of any class is the class of all those classes which are similar to the given class. This definition seems to satisfy the criterion of assigning numbers a single constant meaning, because the only property that such a class of classes appears to have is that of a certain manyness. We shall see later, however, that this appearance does not seem to be entirely correct.

For the sake of precision, let us summarize the most important features of the Russell theory. First and most importantly, natural numbers are regarded as originating in the manyness property of classes rather than the ordering property of transitive-asymmetrical relations. The problem of defining the natural numbers, therefore, becomes synonymous with the problem of defining cardinal numbers. Russell reduced the latter problem to two simpler questions: (1) What does it mean to assert that two classes have the same number? (2) What does it mean to assert that two classes have different numbers? Both (1) and (2) are answered by appealing to relation of correspondence. Two classes have the same number of elements when these elements can be correlated one for one. Two classes have different numbers of elements when their elements cannot be correlated one for one. Because classes have properties other than the manyness of their respective elements, it is not logically proper to regard a number simply as a property (manyness) of classes. It is necessary, instead, to regard each number as a certain class of classes—that is, a class of similar classes. Hence, the series of natural numbers consists of the series of classes of similar classes. Russell summarized this entire definition in just two brief statements: "*The number of a class is the class of all those classes that are similar to it. . . . A number is anything which is the number of some class*" (1919: 18-19).

DISCUSSION

Shortly after Russell's version of the cardinal theory first appeared, several objections to it were raised by idealist philosophers. Many of these may be found in Chapter 2 of Ernst Cassirer's book *Substanzbegriff und Funktionsbegriff* (*Substance and Function*) (1910). In fact, this chapter was written as an idealist reply to the Russell theory. Three general criticisms of a philosophical and psychological nature are posed. First, it seems that the Russell theory is rooted in the ancient and discredited principle of abstraction, whereby a concept (in this case, number) is always defined by abstracting a common property from specific instances of the concept. Cassirer argued that the principle of abstraction implies that numbers are properties of physical objects (with this Russell would have agreed) when, in fact, they are mathematical entities with no direct counterparts in the real world: "The characteristic relations which prevail in the series of numbers cannot be conceived as properties of the given contents of presentation" (1910: 33). The general view that it is inappropriate to regard mathematical and logical constructs as evolving from everyday experience is a standard idealist thesis inspired by Kant but which, as we saw earlier, has been around since Pythagoras. Second, the classificatory approach to numbers was said to be based on an entirely erroneous psychological theory of number. Here, a "psychological theory" means any hypothesis about the function served by numerical ideas in human thinking. Psychologically speaking, Cassirer argued, the principal function of numbers is not, as Russell supposed, that of bringing different things together under a common heading. Instead, numbers serve primarily to keep things separate and distinct: "Number is called universal not because it is contained as a constant property in every individual, but because it represents a constant condition of judgment concerning every individual qua individual" (ibid.: 34). Third, the definition of "number as class" is unacceptably "atomistic" in the sense that it focuses attention exclusively on the individual members of the series of numbers. Cassirer contended that, insofar as mathematics is concerned, we need to ascertain what things are true of the series as a whole. And what is true of the series as a whole is that its elements comprise a progression: "The individual number never means anything by itself alone . . . a constant value is only assigned to it by its position in a total system (ibid.: 48). This, of course, is another example of the old philosophical debate between holism and atomism which dates back to Greek philosophy. Idealist philosophers, as strident opponents of anything even remotely suggestive of analysis, always take the holistic side. In place of Russell's theory, Cassirer, and idealists generally, prefer the Dedekind version of the ordinal theory.

Readers who are especially interested in criticisms of the Russell theory from an idealist philosophical perspective are directed to Cassirer's book and to a much later work, Mathematical Epistemology and Psychology (1966), coauthored by the Dutch mathematician E. W. Beth and the Swiss philosopher-psychologist Jean Piaget. We shall ignore such criticisms, however, because, whatever may be their other merits, they are philosophical in origin. They clearly are predicated on philosophical differences of opinion rather than empirical or logical fact and, worse, they do not raise questions about the logical adequacy of the view that numbers may be defined as classes. Owing to their intrinsic indeterminacy, differences of philosophical opinion fall outside the scope of our inquiry. We shall, instead, take up objections which go to the heart of the Russell theory and suggest certain logical shortcomings.

We shall consider three general objections. First, it has sometimes been contended that the Russell theory is tautological, that is, it does not say anything more than "numbers are numbers." Second, it can also be argued that defining numbers as classes of similar classes permits us to determine whether or not two classes contain the same or different numbers of terms, but it does not allow us to say what these numbers are. Third, it can be argued that while the concept of similarity suffices to define the notion of "same number," the concept of dissimilarity falls short of being an adequate definition of "different number." The first of these criticisms can, it seems, be satisfactorily rebutted. However, the second and third criticisms appear to be unanswerable.

It appears that the tautology criticism was first raised by Cassirer (1910). Essentially, the criticism says that Russell's definition is circular: numbers can be defined as classes only if the concept of number is smuggled into the definition implicitly. This criticism, if it happens to be true, would be absolutely devastating. Two aspects of the Russell theory are suggestive of circularity. First, there is the fact that the theory ultimately comes down to Russell's statement, quoted above, that "a number is anything which is the number of some class." It must be admitted that this statement certainly sounds as though it is circular. The term "number" appears in both the definiens and the definiendum of the definition. Second, the fact that the theory reduces the question, "What are numbers?" to the questions, "What does it mean to say that two collections have the same number?" and

"What does it mean to say that two collections have different numbers?" also seems to entail a circular definition. The term "number" appears in the second and third questions as well as the first. Hence, it looks as though the theory answers the question "What are numbers?" by saying that number means having the same number or having different numbers. But again, the suggestion of circularity is very strong.

Although the Russell theory certainly seems prima facie circular, it actually is not. In the case of the statement that a number is the number of some class, circularity is avoided because the term "number" can be removed from the phrase "the number of some class." Explicitly, "the number of some class" may be defined as the class of all those classes similar to that class. Similarity, in turn, is defined by the relation of one-to-one correspondence. Thus, the prima facie tautological statement that numbers are the numbers of classes is simply a shorthand version of the statement that numbers are classes of similar classes. The same thing is true of the apparent circularity involved in reducing "what are numbers?" to the concepts of same number and different number. The term "number" in both "same number" and "different number" may be defined as "manyness." The phrase "same number" is a shorthand version of "same manyness" and "equally many." Likewise, "different number" is a shorthand version of "different manyness" and "unequally many." The latter concepts may be defined by appealing to the correspondence relation without any reference to number per se. For these reasons, it seems incorrect to charge that the classificatory definition of number is tautological. It only seems so at first glance.

The second criticism may be called numerical discriminability. There is some evidence that Russell was aware of this criticism when he wrote the Principles (1903: 114-15). However, the problem is not explicitly dealt with anywhere in the book, and, in later versions of the theory, it is not even mentioned. Although some illustration will be required to show that there is actually a problem, the general objections are easily stated. First, the similarity relation can perhaps tell us that two classes have the same number of terms, but it cannot tell us what this number is. Also, the dissimilarity relation can perhaps tell us that two classes have different numbers of terms, but it cannot tell us what the numbers are.

Some very elementary illustrations will suffice to demonstrate these claims. Considering the similarity relation, suppose I attend a basketball game with a primitive or with a child who knows no arithmetic. The members of the respective basketball teams wear different colored uniforms—green and white. As the contestants take their positions at center court preparatory to beginning play, suppose my companion discovers that there is one green-clad contestant for each white-clad contestant, and conversely. This discovery would be reasonably likely to occur because, at the beginning of a basketball game, opposing players usually shake hands and stand in close physical proximity to each other. Having determined that there is a one-to-one correspondence between green-clad and white-clad contestants, my companion may conclude that the respective teams are playing with the same number of contestants. But it cannot be concluded that the number of each team is 5. The latter point is made apparent by the fact that if we happened to be attending a women's basketball game, the number in question would be 6 rather than 5. The only possible way to know the specific number of players on each team is to smuggle in some piece of information other than the similarity relation (prior knowledge of the rules of basketball plus a knowledge of gender, counting all the players on the court and dividing by 2, and so on). In short, although similarity suffices to determine that two classes have the same number of terms, the particular number of terms that two similar classes have remains unknown. Strictly speaking, the number could be any finite cardinal. Instead of giving us a particular class of classes, therefore, the similarity relation gives us all classes of similar classes.

The same general problem obtains for dissimilarity. Suppose that my companion is seated in front of two television monitors. On the left-hand monitor, an American collegiate football game is being transmitted and, on the right-hand monitor, a Canadian collegiate football game. Before each play, the offensive team on the left-hand monitor huddles. The offensive team on the right-hand monitor also huddles before each play. At one point during the proceedings, having become thoroughly bored with both games, my companion begins to compare what is occuring on the two monitors. Among other things, the fact that the correspondence between the players in the two offensive huddles is never one-one becomes apparent. From this, it is concluded that the American and Canadian teams are playing with different numbers of contestants. However, apart from counting the players or knowing that American rules provide for 11 players while Canadian rules provide for 12, there is absolutely no way of knowing what the two numbers are. As long as no information other than dissimilarity is available, the actual numbers of contestants on American and Canadian football teams would remain a mystery. They might be any two nonidentical cardinal numbers.

The numerical-discriminability criticism seems especially significant when viewed in terms of what the classificatory approach to numbers originally was supposed to accomplish. It will be recalled that the sole motivation for this theory was the fact that the ordinal theory does not restrict the natural numbers to a single meaning. Therefore, it seems reasonable to judge the cardinal theory in light of this proposed restriction. From what we have just seen, the theory does not appear sufficiently restrictive. The similarity relation provides a basis for inferring numerical equality, and the dissimilarity relation provides a basis for inferring numerical inequality, but neither provides any hints about how to determine what the specific number of any class is. Thus, for example, similarity offers no grounds whereby we might say that pairs of human beings, pairs of shoes, pairs of aces, and so forth, are all instances of the number 2. In so far as similarity is concerned, they might be instances of the number 100. It is to be noted that this is more or less the same criticism that Russell leveled at Peano's version of the ordinal theory. From the standpoint of the original aims of the cardinal theory, therefore, the problem comes down to this. The ordinal theory is dismissed on the ground that it assigns to each natural number more than a single interpretation. We wish to narrow the range of possible interpretations of each natural number to exactly one. To do this, we identify the natural numbers with the property of manyness, and we adopt the general view that numbers are certain types of classes. However, we do not seem to have narrowed the range of interpretations to exactly one. In so far as similarity is concerned, the number of two similar classes can be any number whatsoever.

The third and final criticism which we shall consider is concerned with the completeness of the relation of similarity with respect to the notion of "same number." While it seems obvious that the similarity relation is perfectly sufficient to insure that two (or more) classes have the same number, it is not correspondingly clear that the relationship is necessary to numerical equality. This is because the relation of dissimilarity does not, contrary to Russell's claims, guarantee that two classes have unequally many teams. The concept of dissimilarity, as we saw earlier, means simply that the correspondence between the members of two classes is something other than one-to-one. Given two classes R_x and R_y between whose elements something other than one-to-one correspondence obtains, we know from Chapter 2 that three possibilities remain: (1) the correspondence is one-to-many; (2) the correspondence is many-to-one; (3) the correspondence is many-to-many. When either (1) or (2) is true, this fact is sufficient to insure that R_x and R_y have different numbers. If (1), then the number of the class R_x is less than the number of the class R_y. If (2), then the number of the class R_x is greater than the number of the class R_y. However, if the relation between the members of R_x and R_y is many-to-many, then it is not clear that the two classes have different numbers. For example, consider the following illustration: Suppose that there is a group of exactly 12 authors who have written exactly 12 books. Suppose that each book is coauthored and each author has coauthored more than one book. Each member of the class of authors may be correlated with more than one of the books. Conversely, each book may be correlated with more than one of the authors. Insofar as the notion of dissimilarity depends only on the correspondence relation, the class of authors and the class of books are dissimilar. However, the number of both classes is 12. Obviously, homely examples of this sort could be multiplied indefinitely. For our purposes, this one suffices to establish that the notion of similarity is not synonymous with the notion of "same number." Similarity appears to be a sufficient but not necessary criterion of numerical equality. While similar classes must always have the same number, classes that have the same number are not always similar. It will be recalled that similarity was explicitly proposed by Russell as an answer to the question, "What does it mean to say that two classes have the same number?" As an answer, similarity is not complete because it tells us only part of what it means to say that two classes have the same number. As an answer to the question, "What does it mean to say that two classes have different numbers?", dissimilarity is entirely unsatisfactory. Dissimilar classes may or may not have the same number.

To sum up, we have reviewed three criticisms of the theory of number-as-class in this section. Although it is possible to answer the tautology criticism, satisfactory replies to the last two criticisms do not seem possible without first altering the theory itself in some manner. Evidently, there is nothing in either the definition of similarity or the definition of dissimilarity which may be construed as providing a means of deciding of what number a given class is an instance. By predicating these two notions exclusively on the correspondence relation, we are left with the thesis that numbers, generally, are classes of classes and not much else. The specific number which two classes are cases of appears to be completely indeterminant. Moreover, the third criticism indicates that the similarity relationship is inadequate by itself for even the restricted task of defining what it means to say that two classes have the same number. It would seem that classes other than similar ones may have the same number. This fact suggests that the correspondence relation must not be regarded as dealing solely with numerical (manyness) relationships among classes. By making the correspondence relation the only basis for his definitions of "same number" and "different number," Russell obviously was presupposing, albeit implicitly, that correspondence is a purely numerical relation. Although there may be some way to get around this assumption without altering the cardinal theory, the assumption must be viewed for the present as a flaw in the theory.

5

PIAGET'S AD HOC APPROACH TO NUMBER

> Finite numbers are therefore necessarily at the same time cardinal and ordinal, since it is of the nature of number to be both a system of classes and of asymmetrical relations blended into one operational whole.
>
> Jean Piaget

We turn now to the last of our three approaches to the logical foundations of the number concept. This viewpoint appears to be entirely unique to the Swiss philosopher-psychologist Jean Piaget and his collaborators at the Institute for Genetic Epistemology in Geneva. For reasons that will concern us presently, Piaget's doctrine is not generally regarded by mathematicians and logicians as a viable alternative to the theories we already have considered. Piaget's main thesis is that we should combine the relational and classificatory views of number. It is incorrect, he argues, to base the natural number system exclusively on either ordinal numbers or cardinal numbers. Instead, natural numbers should be identified with both the ordinals and the cardinals. This thesis is predicated on Piaget's belief that, individually, each of the two theories discussed in chapters 3 and 4 tells us only half the story of 1, 2, 3, . . . Whether or not this belief is correct is, as we shall see, open to serious doubt.

Before discussing Piaget's theory further, it is necessary to resolve a question which, though minor from the standpoint of our inquiry, may pose a problem for some readers. The question is concerned with the legitimacy of treating Piaget's number doctrines as though they comprise a logical theory of the same general sort as the Dedekind-Peano and Frege-Russell theories. The question arises because Piaget's views were originally formulated via empirical rather than logical means. Explicitly, they derive from some investigations of the emergence of numerical ideas in children's reasoning conducted during the 1930s. A comprehensive report appeared in 1941 under the title La genèse du nombre chez l'enfant (The Development of Number in the Child). Later, an abridged English translation appeared under the title The Child's Conception of Number (1952). This particular book is by far the most influential and widely read of Piaget's many works.

Piaget presents a psychological theory of number development and empirical findings which are believed to support it. We shall take up both the psychological theory and the ostensibly supportive data in Part II. For the present, however, we must decide whether it is proper to consider the theory as though it also is a logical doctrine. Given the empirical origins of the theory, it may be quite reasonably argued—indeed, it has been argued—that it is improper to consider the theory in this manner. Piaget, so this argument goes, has proposed a purely psychological account of number. Moreover, the argument continues, he wishes the theory to stand or fall entirely on the hard facts of number development in children and not on logical arguments. Generally speaking, it is certainly unwise to mix psychology and logic indiscriminately. Further, it would simplify our task considerably if we did not have to examine the logical grounds for Piaget's theory. We must, however, because the aforementioned argument overlooks two noteworthy points. First, Piaget, in some of his recent writings (Beth and Piaget 1966, Piaget 1970), has discussed his own hypotheses about number primarily in light of certain logical criticisms of Russell's version of the cardinal theory. This certainly seems to suggest that he believes that these hypotheses have logical merit. Second, and more important in my opinion, there is Piaget's underlying philosophy of genetic epistemology. According to genetic epistemology, there is no such thing as a purely logical or psychological theory because logical theories make tacit psychological commitments and psychological theories make tacit logical commitments. To see where this leads us vis-à-vis the question of whether we should consider the logical grounds for Piaget's views, some further remarks about genetic epistemology are necessary.

Normally, a rather firm line is drawn between logic and psychology. This is done to avoid the twin pitfalls of "logicism," whereby psychological facts are inferred from logical arguments, and "psychologism," whereby logical problems are solved by appealing to psychological data. The former is not to be confused with the thesis that all mathematics may be reduced to logic. Logicism leads inexorably to the absurdities of numerology. Psychologism leads to the fallacy of evaluating logical formalizations of concepts in terms of whether they square with everyday usage rather than in terms of whether certain theorems follow from them. Piaget, it must be admitted, rejects both logicism and psychologism in their extreme forms (Beth and Piaget 1966), although it is frequently supposed that he is guilty of the latter. This disclaimer notwithstanding, Piaget maintains that the usual boundary between psychology and logic is unacceptably rigid. Assuming that we bear in mind the mutual independence of their respective methods, he argues that there should be a more complete recognition of the important interactions between logic and psychology.

Toward this end, Piaget adopts a position somewhere between the traditional rigid separation of the disciplines, on the one hand, and the extremes of logicism and psychologism, on the other. He calls his position genetic epistemology, and its key features are these: Although logic and psychology employ quite different methods, there are two inescapable links between them. First, psychology provides the raw material with which logic works. More explicitly, whenever one constructs a logical argument showing that some concept B can be formally defined in terms of some more basic idea A, the very existence of such an argument presupposes that one already grasps A in a psychological sense. That is, one does not "invent" A as a result of logically analyzing B. Rather, A is already part of one's cognitive apparatus, and analyzing B eventually leads us to "discover" the logical connection between B and A. On this point, Piaget favors an illustration which concerns Georg Cantor's development of the theory of transfinite numbers. Cantor's theory, like the Frege-Russell theory of number, is predicated on the correspondence relation. Cantor noted that it is possible to establish a one-to-one correspondence between the series of natural numbers and the series of even natural numbers as follows:

$$\begin{array}{ccccc} 1 & 2 & 3 & 4 & \dots \\ \downarrow & \downarrow & \downarrow & \downarrow & \\ 2 & 4 & 6 & 8 & \dots \end{array}$$

To each element of the upper series, there obviously corresponds one and only one element of the lower series, and conversely. For any given term, x, of the upper series, its corresponding element in the lower series is 2x. For any given term, y, in the lower series, its corresponding element in the upper series is y/2. From this seemingly inconsequential correspondence, a very interesting result followed: a number, which Cantor called aleph-zero after the first letter of the Hebrew alphabet, that is neither an even natural number nor a natural number at all. Piaget argues that Cantor did not, as is commonly supposed, invent his theory from scratch through the application of logical methods to the problem at hand. Although Cantor probably was unaware of it, a grasp of the correspondence relation already was present in his mind, and he simply discovered the connection between it and the theory of transfinite numbers. Piaget explicitly maintains regarding Cantor's use of the correspondence relation that "He found it in his own thinking; it had already been part of his mental equipment long before he even turned to mathematics" (1970: 5).

Therefore, the first interface between logic and psychology is that psychology constrains the content of logic. By virtue of logical methods, we discover that some concepts which we already possess logically entail other concepts which we already possess, but we do not invent entirely new concepts via these methods. If this sounds vaguely like the Pythagorean and Platonic doctrines about mathematical ideas considered in Chapter 1, that is only because it is. The second interface between logic and psychology is that "there is parallelism between the progress made in the logical and rational organization of knowledge and the corresponding formative psychological processes" (ibid. : 13), which occurs because logical theories make tacit psychological commitments. Whenever we construct an ostensibly logical theory, Piaget argues, we invariably refer to psychological data, usually without knowing it, and we thereby commit ourselves to certain psychological assumptions. It is not certain whether this argument is true in general, at least not without more extensive empirical evidence than is currently available. However, it is not difficult to find specific cases in which the argument undoubtedly is true. For example, recall Russell's development of the theory that numbers are classes. We have seen that the point of departure for this theory is an essentially psychological assumption, that is, that in everyday life number is chiefly concerned with experiences of manyness and correspondence rather than with experiences of ordering. Russell also makes the historical assumption that number originally became important to human civilization as an abstraction of the manyness property of classes (1919). Whatever may be the merits of these assumptions, it must be admitted that they are matters of empirical fact and not of logic.

But let us return to the supposed parallelism between logical theories and the "formative psychological processes." In _Mathematical Epistemology and Psychology_, we are further enlightened on the nature of this parallelism:

> As against Pasch, who believed that 'mathematical thought advances in opposition to human nature,' all that we have learned from the study of the development of intelligence forces us, on the contrary, to believe that the transcending of empirical and even of pure intuition by the refinement of deductive methods, artificial as they may sometimes seem, is inherent in the 'natural' continuation of many other kinds of such transcendence. Pasch's illusion concerning 'human nature' arises simply from the fact that, like so many writers, he has judged it by too brief observations of adults other than himself or by incomplete introspection. If he had been in possession, as we are today, of certain data

> on the transformations of logico-mathematical activity between the first and the fifteenth year of mental development, perhaps he would have appreciated that the axiomatician Pasch is much more in the line of such a development than is his 'human nature,' as he represented it without being sufficiently aware of the profound laws of his own genetic development. . . . It follows that the line of thought which substitutes hypothetico-deductive procedures for operational intuition is already built into the line of development at relatively elementary stages, and that the reversal of perspectives which ends up with axiomatic reorganization and with formalization, is not opposed to nature but on the contrary appears as 'natural' as the pre-axiomatic constructions (Beth and Piaget 1966: 134; emphasis added).

Decoding this decidedly opaque passage is something of a problem in cryptography. Beneath all the obscure phrases, however, there is a fairly simple message. We know genetic epistemology says that the methods of logic treat concepts that are already part of our mental equipment. By virtue of such methods, we demonstrate that some concepts (for example, the concepts of arithmetic) logically imply other concepts (the concepts of algebra). This is what "axiomatic reorganization and formalization" refers to. Now, suppose we have a logical theory which says that some concept or concepts, B, follows from some second concept or concepts, A. Because psychology constrains logic, B and A must emerge naturally during the course of spontaneous cognitive growth. More important, they must emerge in a fixed order such that A precedes B. Regarding this latter prediction, Piaget frequently appeals to a geometric illustration. Historically, Euclidean geometry evolved much earlier than topology—about 23 centuries earlier to be precise. But logical analysis shows that Euclidean geometry follows from topology; that is, the theorems of Euclidean geometry are true if the theorems of topology are true but not conversely. If there was no "parallelism" between logic and psychology, it might be supposed that the spontaneous development of geometric concepts in children would reflect the historical order of emergence with Euclidean ideas appearing first. However, Piaget reports that spontaneous cognitive development follows the logical sequence rather than the historical sequence. Data reported in *The Child's Conception of Space*, for example, seem to indicate that topological concepts such as "closed" and "open" appear during the preschool years while the Euclidean concepts which comprise the core of most high school geometry courses do not appear until the late elementary school years.

Therefore, the second interface between logic and psychology is that the logical and developmental orders of things are the same. If, by virtue of logical methods, we discover that some concept A implies some other concept B, then B must follow A developmentally. If developmental research fails to confirm this prediction, then, assuming we can rule out the possibility of having committed measurement errors in our research, the logical theory probably is incorrect. This leads Piaget to conclude that it is crucial for logical theories to be supplemented by developmental data. What seems to be suggested is a three-step process in which the methods of logic and psychology remain distinct but their resulting data interact. First, after making certain implicit or explicit psychological assumptions, we construct a logical theory of some sort that must satisfy strictly logical criteria. This theory leads to certain predictions about the spontaneous development of the concepts with which it deals. The second step consists of testing these predictions; if the predictions are not confirmed, then we have grounds for doubting the original assumptions on which the logical theory was based. Third, after adjusting our psychological assumptions to take the data into account, we construct a new logical theory. Although this must be done via exclusively logical methods, the developmental data may provide some hints about where to begin.

If one takes Piaget's epistemology seriously, it seems impossible to escape the conclusion that there is no such thing as a purely logical theory. Such a theory always has psychological implications, which are subject to empirical testing and which have implications for the theory itself. Therefore, it would be inconsistent with genetic epistemology to argue that any of Piaget's theories about the spontaneous development of concepts—as long as the concepts in question are ones which sometimes appear in logical theories—are "purely psychological" and are intended to stand or fall solely on empirical data. Although implausible from the standpoint of genetic epistemology, this argument does have the advantage of convenience, which no doubt is responsible for such popularity as it may enjoy. However, this chapter proposes to consider the logical grounds for Piaget's views. Moreover, even if the aforementioned reasons for considering them did not exist, there would be nothing intrinsically improper in doing so anyway. Independent of Piaget's own position, there is positive knowledge to be gained from determining whether or not the thesis that natural numbers are equal parts ordinal and cardinal has logical merit. This fact alone suffices to justify examining the thesis.

THE THEORY

From a logical standpoint, Piaget's approach may be regarded as an attempt to answer certain criticisms lodged against Russell's version of the cardinal theory by intuitionists such as L. E. J. Brouwer and Henri Poincaré. These criticisms are psychological in nature, which is why we did not examine them in the preceding chapter. Intuitionists, it will be recalled, do not believe that mathematics either can be reduced to logic without loss of meaning or is simply a game played with otherwise meaningless symbols. They believe mathematics treats meaningful concepts which are not purely logical. Alfred Heyting once summarized this general outlook as follows: "The intuitionist mathematician proposes to do mathematics as a natural function of his intellect, as a free, vital activity of thought. For him, mathematics is a production of the human mind. He uses language, both natural and formalized, only for communicating thoughts, i.e., to get others or himself to follow his own mathematical ideas" (Benacerraf and Putnam 1964: 42).

It should be obvious from this passage that, in the intuitionist scheme, the science which underlies mathematics is psychology rather than logic. Insofar as schools of psychology are concerned, intuitionism, like modern linguistics, leans toward nativism. That is, the fundamental concepts with which mathematics deals are, like the fact that we each have two eyes and one nose, the result of species-specific hereditary programming. Simply put, the building blocks of mathematics are given by heredity and mathematics proceeds from there. Concerning the number concept in particular, Brouwer contended that even infants have a native "intuition" of the unending sequence of natural numbers. Similarly, Poincaré contended that the rule that generates the arithmetical progression to which the natural numbers belong (add 1 to a given term to obtain the next term) is, somehow, innately given. Most psychologists, given what they know about the complexities of intellectual growth in children, would regard these claims as fantastic oversimplifications. The fact that intuitionism continues to be an important influence in modern mathematics suggests that mathematicians are far less sceptical about such claims than psychologists are.

Although Piaget believed that something was wrong with the Russell theory and that intuitionists were correct in objecting to it, he did not agree with their specific objections. In particular, he parted company with them on the question of innateness. Piaget reasoned that the view that number is a natively given intuition entails that an understanding of the natural number system must appear full-blown at some point during development. (It is not at all clear that this belief, which is based on a very narrow conception of nativism, is correct.) Piaget's own investigations of children's numerical ideas did not confirm this prediction (Piaget & Szeminska 1941). They showed a protracted period of numerical development spanning, roughly, the late preschool years and the first year or so of elementary school. From this fact, Piaget concluded that the number concept cannot possibly be innate. He also concluded that it is correct to view mathematics, generally, and number, particularly, as reducible to logical concepts. But he continued to believe that there was something logically incorrect about saying that numbers are merely classes of similar classes. Like Cassirer, he decided that this definition must be tautological:

> Russell and Whitehead have not used the qualified one-to-one correspondence that is used in classification. They have used the correspondence in which the elements become unities. They are, therefore, not basing number only on classification operations as they intend. They have, in fact, got themselves into a vicious circle, because they are attempting to build the notion of number on the basis of one-to-one correspondence, but in order to establish this correspondence they have been obliged to call upon an arithmetic unity, that is, to introduce the notion of a nonqualified element and numerical unity in order to carry out the one-to-one correspondence. In order to construct numbers from classes, they have introduced numbers into classes (1970: 37).

Once again, Piaget's language and reasoning are obscure, but his point is elementary: Defining a number as the number of some class is circular. At the end of the preceding chapter, we saw that the apparent circularity of this definition is mitigated by the fact that "number of some class" can be defined by appealing to the similarity relation. The above passage implies that this is insufficient. Piaget believes that the notion of one-to-one correspondence entails not only that the terms of two classes are equally populous but also that to each specific term in one class there corresponds a unique term in the other class, and conversely. Russell's definition fails on this point:

> When the months of the year, the apostles of Christ, Napoleon's marshals and the signs of the zodiac are made to correspond so that the number 12 as a class of these classes is abstracted from them, this is not because there is a qualitative equivalence between the month of February, the apostle Peter, the marshall Ney and the sign of Cancer, but because any element of one of these classes can be made to correspond to any of the others independently of their qualities (Beth and Piaget 1966: 264).

Without such a "qualitative equivalence," one-to-one correspondence is impossible. The only way that Russell was able to effect such correspondences, so the argument goes, is by assuming, without knowing it, that he knew the number of which given classes were instances. This is an exceedingly elliptical argument in which there appears to be no merit whatsoever. First, Piaget seems to confuse the everyday "concrete operation" of correspondence with the formal relational concept which we examined in Chapter 2. While it may be true that, in our everyday use of correspondence, we require some equivalence relationship between individual element pairs to be certain that two classes have the same number, it is not true that in logic we must know that, for example, Napoleon always corresponds to Josephine as a precondition for concluding that there is a one-to-one correspondence between the class of monogamously wedded men and the class of monogamously wedded women. The relation of monogamy suffices to guarantee the correspondence independent of links between specific members of the two classes. Second, although the Russell definition obviously would be circular if Russell had "introduced numbers into classes," there is nothing in our earlier synopsis of the theory, or in Russell's three versions of the theory, which suggest he did this. For the sake of discussion, however, let us suppose that Piaget's point about the impossibility of establishing one-to-one correspondences between classes of unqualified elements is correct and see where it leads him.

As it turns out, it leads him back to order or, more generally, transitive-asymmetrical relations. Piaget's argument requires some method, in addition to correspondence, whereby we can map an otherwise unqualified element of one class with one and only one otherwise unqualified element of another class, and conversely. Given that we desire a logical theory of number, the method must not be empirical but, rather, must be completely definable in terms of the primitive concepts of logic. The method also must be equally applicable to all classes. Thus, in the case of the class of monogamously wedded men and the class of monogamously wedded women, an exhaustive catalogue, reporting who has married whom, is unsatisfactory both because it employs empirical data and because it offers no general principle for other pairs of classes. The method that Piaget ultimately decided on is that of ordering the terms in classes. If we are allowed to order the respective terms of two classes R_x and R_y, then we may say that R_x has a first term x_1, a second term x_2, and so on, and we may say that R_y has a first term y_1, a second term y_2, and so on. If R_x and R_y happen to be finite, then we may also say that each has a last term. Once the terms in each class are ordered, we can map x_1 with y_1, x_2 with y_2, and so on. The ordering method satisfies Piaget's argument because it provides a property or, in Piaget's words, a "qualification" whereby each term in one class can be mapped with a unique term in another class and conversely—the property of ordinal number. Two terms are mapped together if and only if they have the same ordinal number. This method is "logical" because, as we saw earlier, ordinal number can be satisfactorily defined in terms of the concepts of the logic of relations. Moreover, the method can be applied to all classes, finite or infinite, without exception.

Piaget summarizes the case for introducing an ordering of terms into the Russell theory by observing that

> along with the classifying structures . . . number is also based on ordering structures. . . . It is certainly true that classification is involved in the notion of number. . . . But we also need order relationships for this reason: if we consider the elements of the class to be equivalent (and this of course is the basis of the notion of number), then by this fact it is impossible to distinguish one element from another—it is impossible to tell the elements apart. We get the tautology A + A = A; we have a logical tautology instead of a numerical series. Given all these elements, then, whose distinctive qualities we are ignoring, how are we going to distinguish them? The only possible way is to introduce some order. . . . This relationship of order is the only way in which elements, which are otherwise being considered as identical, can be distinguished from one another (1970a: 38).

To summarize, Piaget's views about the logical foundations of the number concept may be thought of as consisting of a few simple statements. First, Piaget obviously assumes that Russell was correct when he maintained that the natural numbers should be identified with the cardinal numbers. The importance of this initial fact can hardly be overemphasized. We saw in the preceding chapter that there are reasons for doubting that the connection between number and the manyness property of classes is as intimate as the Frege-Russell theory supposes. Piaget apparently does not countenance these arguments. Second, although the number concept derives from the notion of correspondence between classes, this is not all it involves. If our theory does not provide for some relation in addition to correspondence, then we shall not be able to establish the one-to-one correspondences which are necessary to avoid circularity. Third, in addition to correspondence, we should introduce order as a primitive notion in our theory. This gives each term in any class an ordinal number. By employing the rule that paired terms from different classes must have the same ordinal number, we insure that, given two classes with equally many terms, each term in one class is paired with one and only one term in the other class, and conversely. Parenthetically, it should be noted that Piaget's approach would eliminate the numerical indeterminacy of many-to-many correspondences discussed at the end of the preceding chapter.

DISCUSSION

There are many logical objections to the view that natural numbers should be regarded as simultaneously ordinal and cardinal. With the exception of the numerical indeterminacy of many-to-many correspondences, this view does not elminiate all of the criticisms of the ordinal and cardinal theories discussed in the preceding two chapters. Moreover, it raises some entirely new questions. We will consider two such questions; both appear to be reasonably conclusive grounds for rejecting the theory.

The first criticism is concerned with Piaget's thesis that the only way to insure one-to-one correspondence between classes containing equally many terms is to allow the elements in the respective classes to be ordered. Against this thesis, two points may be raised. First, we know that Piaget wishes to insure that, for any two classes which contain equally many elements, we can specify precisely which element of one class is paired with a given element from the other class. A single counterexample, based on an earlier illustration, suffices to show that it is not always necessary to introduce order into classes to achieve this result. Recall Cantor's one-one correspondence between the natural numbers and the even natural numbers. Earlier, this correspondence was introduced by showing the first four elements of each class arranged in their usual order. However, the correspondence does not depend in any way on this order, or any order for that matter. Suppose we have the class of all natural numbers and the class of all even natural numbers, but neither is arranged in any particular order. This does not prevent us from establishing a one-to-one correspondence between them. Given any natural number x, we do not need to know its position (ordinal number) in the series to find its even correspondent. We simply multiply by 2. Similarly, given any even number y, we do not need to know its ordinal number to find its natural correspondent. We simply divide by 2. Given any term from either class, therefore, we can map it with a unique term in the other class without knowing the ordinal number of either term. Hence, the one-to-one correspondence between the natural numbers and the even natural numbers does not appear to turn on the ordering of the terms. What it does depend on is the fact that the class of natural numbers contains an infinitely many members. Mathematical logicians have proved a very important theorem about such classes which many readers will no doubt find somewhat paradoxical. The terms in any infinite class can be placed in one-to-one correspondence with the terms of subsets of the class. It is this fact, rather than ordering, which is responsible for a Cantor's one-to-one correspondence. For precisely the same reason, the natural numbers may be placed in one-to-one correspondence with their squares, their cubes, and so on. More generally, given any class R_x which contains infinitely many values in the domain of x that satisfy the propositional function, it is known that these values can be placed in one-to-one correspondence with certain subsets of these values (for proofs, see Dedekind 1887 and Russell 1903). This fact, which has nothing to do with order, is one of the most important theorems of modern set theory.

The second point to be raised about Piaget's thesis concerns whether or not his emphasis on determining what specific element of one class is mapped with what specific element of another class is logically or empirically motivated. To stand as a criticism of the Frege-Russell theory, it must be the former, but it apparently is the latter. It will be recalled that Piaget claims it is impossible to conclude that two classes are, in fact, similar unless we have a method for determining what specific elements are mapped with each other. There do not appear to be any logical grounds for this claim. We saw in Chapter 2 that the nature of the correspondence between two classes does not depend on mapping specific values together. It does not even depend on knowing specific values which satisfy the respective propositional functions. In logic, correspondence is not a mapping "operation" which mechanically links one element with another, as Piaget seems to think it is. Correspondence depends on the relations which obtain between classes. Correspondence is, as we saw in Chapter 2, a "property" of certain kinds of relations. Some relations are invariably one-to-one (for example, monogamy) in the sense that every value which satisfies one unknown term corresponds to one and only one value which satisfies the other unknown term, and conversely. When such a relation is partitioned into classes by converting the propositional function in two unknowns into separate propositional functions in one unknown, the correspondence obviously remains one-one. Similarly, there are some relations which are invariably one-to-many or many-to-one or many-to-many. Converting such relations into propositional functions in one unknown does not alter these facts. We may say that all relations of this sort are "logically determinate" vis-à-vis correspondence, because the nature of the correspondence between the classes which would result from partitioning the relation is determined by the relation itself. Because the cardinal theory is, after all, a logical theory, it is concerned, when it comes to correspondence, with logically determinate relations. However, there are classes between which no logically determinate relation exists, or at least none that we know of. In such cases, correspondence is not a formal property of relations. It is, rather, a purely empirical question. The Frege-Russell theory, or any other logical theory for that matter, is not concerned with such cases. Piaget, however, has chosen to base his thesis on an illustration in which the respective classes are not distinct halves of a logically determinate relation. The fact that, for example, Christ had as many disciples as Napoleon had generals and conversely is simply a historical accident and, for that reason, poses no logical challenge.

A second, and logically more serious, objection to identifying the natural numbers with both ordinal and cardinal numbers is that, in two ways, it is simply nonparsimonious to do so. First, let us recall the overriding aim of any definition of the natural numbers—to make mathematics itself possible. Each of either the ordinal theory discussed in Chapter 3 or the cardinal theory discussed in Chapter 4, by itself, suffices to make the theorems of arithmetic and the rest of mathematics true. The theorems proved in Dedekind's (1887) treatise establish the sufficiency of the ordinal theory as a foundation for mathematics. Similarly, the theorems proved in the first volume of *Principia mathematica* establish the sufficiency of the cardinal theory. This means that, in so far as our principal criterion is concerned, we must choose between the view that natural numbers are values which satisfy the unknown terms of transitive-assymmetrical relations and the view that natural numbers are classes of similar classes. As the concluding sections of the preceding two chapters suggest, there is admittedly some doubt about which approach is the better one. Because both theories satisfy our original aim, our selection ultimately must be based on criteria other than making the rest of mathematics possible. Regardless of which approach we select, however, there appears to be no doubt that we must choose. To do otherwise, as Piaget has done, is to be guilty of superfluity. Russell was well acquainted with this fact, and it led him, in addition to proposing his own theory, to argue for the complete irrelevance of order to number:

> The notion of similarity is logically presupposed in the operation of counting, and is logically simpler though less familiar. In counting, it is necessary to take the objects counted in a certain order . . . but order is not the essence of number: *it is an irrelevant addition,* an unnecessary complication from the logical point of view. The notion of similarity does not demand order . . . (1919: 17; emphasis added).

The second sense in which Piaget's theory is nonparsimonious is that it is logically impossible for both ordinal number and cardinal number to be primitive notions vis-à-vis natural number: A definition of either notion always entails a definition of the other. Thus, only one of them, the one for which the first definition is given, may properly be regarded as primitive. We already saw how cardinals follow from ordinals at the beginning of Chapter 3. Given that we know the ordinal number of a certain term of any progression, then we also know how many terms there are in the progression up to and including the given term. Given that we know "the" ordinal number of a certain progression, then we also know how many terms the progression contains. The ordinals also follow from the cardinals in a somewhat less direct manner. The cardinals, when properly defined, can be shown to imply a definition of the notion of simple infinite progression. (In fact, the cardinals are a simple infinite progression.) Since ordinal numbers are nothing more than the common property of all simple infinite progressions, we have a definition of the ordinals. Russell summarizes the cardinal definition of the ordinals as follows: "We may therefore define the ordinal number n as the class of serial relations whose domains have n terms, where n is a finite cardinal number" (1903: 243). The crucial point is that it is completely superfluous to "add" ordinal number to Russell's definition of cardinal number as Piaget does. If we accept the notion of order as primitive to number, then the correspondence relation—whose sole function is to define cardinal number—becomes redundant. We have a definition of cardinal number from which, if we wish, correspondence can be deduced. Similarly, if we accept the notion of correspondence as primitive to number, then the notion of transitive-asymmetrical relation, whose sole function is to define ordinal number, becomes redundant. We have a definition of cardinal number from which the notions of ordinal number and transitive-asymmetrical relation can be deduced. In all of Piaget's extensive writings on the number concept, there seems to be no evidence that he is aware of either of these facts.

For the reasons we have just considered, Piaget's approach, as a logical theory, cannot be said to be a reasonable alternative to either the Dedekind-Peano or Frege-Russell theories. Although the evidence for the latter theories is far from ineluctable, neither suffers from a fault so basic as superfluity. Whether Piaget's approach is psychologically sound is quite a different question. We shall take up this question and others in Part II.

6

THEORIES OF NUMBER DEVELOPMENT

In addition to the very obvious differences in methodology, the psychological approach to explaining the origins of the number concept differs from the logical approach in three general ways. First, recalling the disagreement over whether numbers were "discovered" or "invented" discussed in Chapter 1, there is no serious division on this point within psychology. With the possible exception of some psychophysicists, such as S. J Rule, there is more or less unanimous agreement among psychologists that the second position (invention) is the correct one. We saw earlier that there are groups of mathematicians who maintain that numbers either are real in a Platonic sense (neo-Pythagoreans) or are coded on the chromosomes (intuitionists). Working psychologists tend to view both of these claims as little more than question begging. The Pythagorean doctrine is metaphysical and, consequently, simply a non sequitur from an empirical standpoint. The intuitionist doctrine, though prima facie empirical, turns out to be almost as unsatisfactory as the Pythagorean. We know so little about what—if any—the connections are between genes and psychological traits that to say that any concept is innate is all but vacuous. In place of Pythagoreanism and intuitionism, psychologists have tended to adopt the view that numerical concepts are learned, usually during childhood, and that sociocultural variables play a vastly more important role in this process than either God or DNA. Two general lines of evidence are believed to substantiate this view: (1) humans reared in cultures radically different from our own (for example, hunting-and-gathering cultures) lack even the most rudimentary numerical skills, and (2) the course of numerical development in children reared in industrialized nations is quite protracted. On the latter point, what we know of number development in Western children suggests that there are few, if any, of the abrupt shifts and leaps which usually characterize the development of capacities under direct hereditary control (for example, secondary sexual characteristics).

A second difference between the logical and psychological approaches concerns their respective goals. We know that the overriding aim of the former is to abstract from 1, 2, 3, . . . those properties from which the truth of the theorems of arithmetic may be shown to follow. If it can be demonstrated that the first laws of arithmetic follow from a given property, we have a logically satisfactory definition. Psychology, on the other hand, is concerned to explain the fact of arithmetic computation. Therefore, a psychological theory of number must at least say what concepts are necessary preconditions for being able to do arithmetic. If a theory posits that some concept A is a necessary precondition for arithmetic, this leads to two empirical predictions. First, there should be a "developmental lag" between A and arithmetic such that children must make considerable prior progress with A. Second, there should be a "functional connection" between A and arithmetic such that increments in A tend to produce increments in arithmetic. Confirmation of predictions of the former sort may be thought of as the necessary empirical criterion of any psychological theory of number. If a given concept does not invariably precede arithmetic in children's thinking, then it cannot possibly be a necessary precondition for arithmetic. However, the mere fact that a concept actually does precede arithmetic does not, in itself, permit us to conclude that the concept plays an important role in the development of arithmetic. Concepts such as "left," "above," and "inside" appear before arithmetic, but no one would suppose that they have very much to do with its development. The latter conclusion follows only from confirmation of predictions of the second sort, which, for this reason, may be thought of as the sufficient empirical criterion for a psychological theory of number.

The third difference between the logical and psychological approaches is implicit in what has just been said. The target phenomenon, the working definition of "natural number," is different in each case. In so far as logic is concerned, the theorems of arithmetic compose the working definition of natural number. In the case of psychology, real behaviors, such as actual arithmetic computation, constitute the working definition. Logic must say why it is that formulas such as $7 + 4 + 2 = 13$ always have certain formal properties. But psychology must say what concept or concepts are prerequisites for solving problems such as $2 + 3 = ?$, $4 - 1 = ?$, $5 + 1 = ?$, and so on. Therefore, psychology normally understands "natural number" to mean the fact of solving the simplest computational problems of arithmetic, while logic understands "natural number" to mean the fact that all arithmetic expressions are associative, commutative, and distributive.* These

*We shall see that Piaget employs a slightly different psychological definition of natural number. As we shall also see, however, he proposes a connection between his definition and arithmetic.

are very important differences in meaning which entail that our operational definition (or criterion) of the number concept in the chapters that follow will be quite different than in Part I.

At present, the psychological literature contains three competing theories about the origins of number concepts. All of them take as their starting points behavioral counterparts of two logical ideas discussed in Part I. The theories are, in the order that we shall examine them, (1) the cardinal theory, (2) the ordinal theory, and (3) Piaget's cardinal-ordinal theory. We shall not review the empirical evidence for and against them. The aim of this particular chapter is simply to report the theories themselves. We shall postpone the data until subsequent chapters.

THE CARDINAL THEORY

This theory is the oldest and, insofar as psychological opinion is concerned, the most redoubtable of our three theories. The theory comes primarily from experimental psychology, and it is especially influential among investigators engaged in the study of psychophysics (for example, Underwood 1966) and the study of learning (Staats 1968). It also has some adherents among developmental psychologists. Generally speaking, the cardinal theory is a straightforward translation of the Frege-Russell theory into psychological terms. It is assumed that, in their simplest sense, numbers refer to classes, and the theory stresses that number ultimately is based on physical counterparts of the logical idea of manyness. The cardinal theory is so influential among experimental psychologists that it is only rarely stated as theory rather than as fact. However, a comprehensive development of the theory was given in a paper by T. M. Nelson and S. H. Bartley (1961) and the present account is based, for the most part, on this excellent paper.

First, and most importantly, Russell's argument that number is basically a method whereby otherwise dissimilar things are gathered together into classes is accepted as psychologically as well as logically correct. Therefore, the task of explaining the psychological origins of number becomes synonymous with explaining how children first come to grasp cardinal number. Concerning this question, the cardinal theory seeks to formulate a property of nature which is both readily perceptible by children and may be used to generate the concept of cardinal number. It is proposed that there is a physical counterpart of the logical notion of manyness, and that this property of stimuli is easily perceived—even by children. "Numerousness," sometimes called "numerosity," is concerned with "perceived differences in manyness." It may be operationally defined by referring to Figure 6.1.

FIGURE 6.1

Illustration of the Usual Operational Definition of Numerousness Employed by Experimental Psychologists.

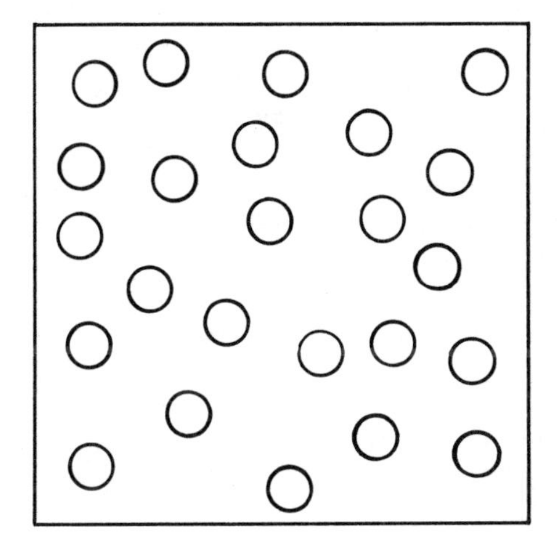

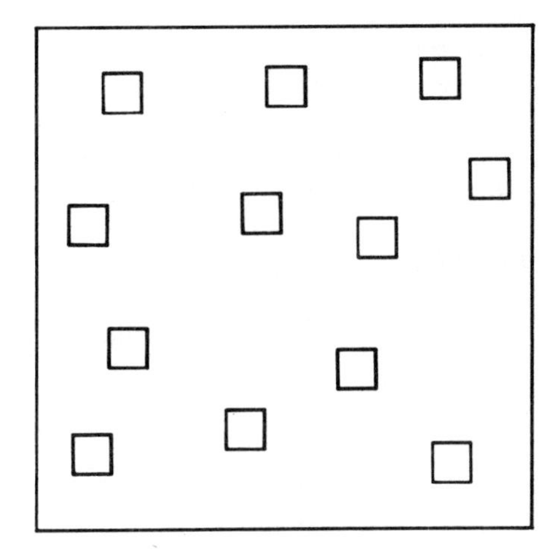

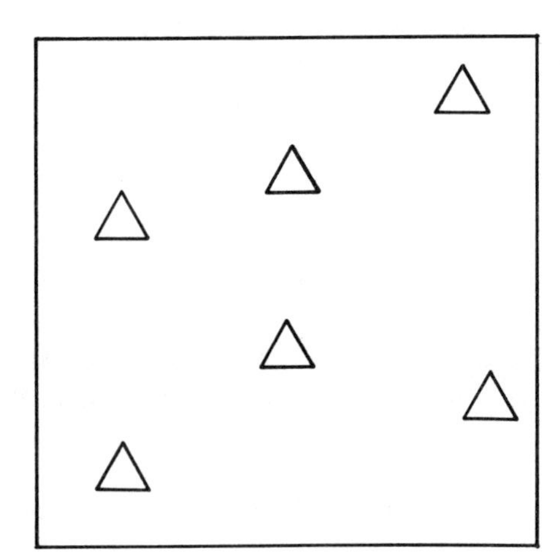

Source: Prepared by the author.

The collection at the top of Figure 6.1 consists of 24 circles, whereas the collections at the bottom consist of 12 squares and 6 triangles, respectively. Suppose an experimenter and a subject are seated across from each other at a table, and suppose that a card on which Figure 6.1 is printed is placed face-down on the table between them. The experimenter turns the card face-up and allows the subject to look at it for a brief interval—say, two seconds. The experimenter then turns the card face-down again and asks a question such as, "Are there more circles than there are triangles?" After the subject has answered, the experimenter turns the card face-up again for another brief exposure, and asks a question such as, "Are there as many triangles as there are circles?" And so on. If the subject can correctly judge the manyness differences between these collections, then we say that he or she perceives numerousness. Note that the subject's judgment must be perceptual rather than conceptual, because the exposure time is too brief for him to count all the objects in the three collections.

A great deal of research on perceived numerosity has been done by psychophysicists. The relationship between judgment accuracy and (1) absolute manyness of collections, (2) stimulus exposure time, and (3) manyness differences between collections has been examined in numerous experiments. Generally speaking, perceived numerosity improves as the absolute manyness of collections decreases, stimulus exposure time increases, and the manyness difference between collections increases (Underwood 1966). More important for our purposes, some studies of perceived numerosity have been conducted with children (for example, Taves 1941). It is known that—as long as the collections to be compared contain 10 elements or fewer, the exposure times are a few seconds in duration, and the manyness difference between the classes is reasonably large—children are capable of perceiving numerosity long before they show any evidence of arithmetic. This fact leads advocates of the cardinal theory to conclude that there are certain instances in which we can directly perceive number without reference to arithmetic; this leads to the further conclusion that classes are not merely abstractions but exist in nature as perceptible entities:

> There exist 'natural classes' of one, two, three, etc. objects, which, if they are not too large, are something perceivable or directly discriminable . . . such natural classes are items that are perceivable even by various subhuman species. . . . Let us say that these natural classes have <u>numerosity</u>. <u>It is numerosity, not number, that is discriminative in nature</u> (Bartley and Nelson 1961: 179).

Thus, there is a property of nature that corresponds to the logical property of manyness. Moreover, it would seem that the first few cardinal numbers are directly perceptible. Both conclusions are predicated on the fact that, under certain conditions, even very young children can perceive numerosity. But how does the number concept evolve from this percept? All that is involved, or so the cardinal theory proposes, is the addition of symbolization. That is, we invent a symbol system the sole task of which is to represent the property of manyness. This gives us a convenient shorthand for the "natural classes" which are the first few cardinal numbers. This shorthand may be used to derive larger cardinal numbers which are not directly perceptible (for example, 2^{14}). The important point is that, psychologically speaking, when we first learn to say "one," "two," "three," . . . or to write "1," "2," "3," . . . , all we are doing is expressing a symbolic convention denoting numerosity: "What is written as a natural number is a conceptualization referring to the natural class of events or things previously denoted as numerosity" (Bartley and Nelson: 180). And numerosity is simply the perceived physical property from which the logical idea of manyness is derived: "Numerosity is a natural fact. It has to do with perceived or inferred manyness" (ibid.).

Thus, we have a psychological characterization of the number concept which says that the first few natural numbers refer to manynesses which may be directly perceived in the physical environment. While the first few natural numbers are "percepts," the larger ones are "concepts" in that they cannot be directly perceived and must be derived from earlier numbers. This leaves us, however, with a question about the way numbers actually are used in arithmetic and higher mathematics. If the natural numbers—which, after all, are the substratum from which mathematics originally evolved—ultimately are only concerned with manyness, then how is it that their manyness property is not important in classical mathematics? We saw in Part I that the truth of the theorems of arithmetic turns on the fact that the natural numbers are arranged in an immutable order and not on the fact that each represents a certain manyness. If, as the cardinal theory supposes, the natural numbers are initially devices for denoting the perceptual counterpart of manyness, then the relative insignificance of manyness in mathematics seems a psychologically improbable result which requires explanation. The proposed explanation is that natural numbers also are abstract symbols which, once they are written down on paper, may have properties other than their ultimate psychological meaning. One of these properties, obviously, is that the symbols always are set down in the same order. Mathematics can make use of "acquired" properties such as this, either in addition to manyness or to the exclusion of manyness:

> Numbers, however, as abstractions, while they may represent numerosity, can and also do differ in an important respect from numerosity. Numbers have an independent existence, as, say, marks on paper, from events with numerosity. They can be treated symbolically as though they are completely divorced from natural numerosities. Studied only as 'conceptual entities' they partake of many qualities that are purely formal. Hence we can have transcendental or imaginary numbers, but not transcendental or imaginary numerosity (Bartley and Nelson 1961: 180).

This does not seem a very satisfying explanation of the fact that mathematics has to do with order and not with manyness. The explanation is ad hoc in that it does not say why, of all the "purely formal" properties which are available, order was selected. The explanation also entails that there are two distinct steps in the development of the number concept. We begin by learning to associate "one," "two," "three," . . . with perceived manyness. Therefore, there is an initial age level, presumably during the preschool years, when "what number" means only "how many." Later, when we begin to make systematic use of the number concept, we learn that "one," "two," "three," . . . also refer to order or position. Thus, there is a second level, presumably during the elementary school years, when "what number" means both "how many" and "what position." Now, for some as yet unknown reason, the latter meaning eventually becomes dominant and, in arithmetic and the rest of higher mathematics, no further mention of manyness occurs. If the number concept actually is rooted in numerosity, then we would say that this is, prima facie, an improbable result. In the absence of a definitive explanation of why this shift in meaning occurs, it seems far more parsimonious to suppose that the number concept is rooted in a perceived physical counterpart of order and that its manyness reference is incidental. It does not seem that those who advocate the cardinal theory have considered the merits of this view.

Although detailed consideration of the evidence for and against our psychological theories of number is being deferred to subsequent chapters, there are certain difficulties associated with the theory we have just reviewed which merit immediate attention. First and most importantly, the operational definition of numerosity, as we described it earlier, tends to confuse the logical idea of manyness with the physical property of "density." Other things being equal, collections containing more terms will be more dense than collections containing fewer terms. As the manyness difference between two classes increases, so does the discrepancy between their respective densities. Certainly this is the case for the collections in Figure 6.1, and it also is the case for the collections employed in numerosity research. Hence, whenever such collections are employed, it is possible that judgments of numerical equivalence and numerical difference may be based on perceived density rather than perceived manyness. If the density cue could be controlled in such a way that correct judgments could only be based on manyness, it might turn out that manyness is not so highly perceptible a property as the cardinal theory supposes. A very elementary method of eliminating the confounding of density with manyness is illustrated in Figure 6. 2. The terms of the collections to be compared are arranged in parallel rows. When two collections do not have equally many terms, as at the top of Figure 6. 2, the smaller collection is the more dense. When two collections have equally many terms, as at the bottom of Figure 6. 2, one of them is more dense. We shall see in Chapter 7 that only the latter method is a completely satisfactory control. We shall also see that judging the relative manyness of two collections becomes a much more difficult matter when perceived density is controlled in this way. Whereas preschoolers routinely make correct judgments after very brief exposure to stimuli in which density and manyness are confounded, even nine- and ten-year-olds have difficulty making such judgments when manyness is the only available cue. At the very least, this seems to suggest that, although density may be highly perceptible property of the physical world, manyness probably is not. If this is true, then the claim that the first few cardinal numbers are directly perceptible "natural classes" seems extremely doubtful.

A second difficulty with the cardinal theory concerns the correspondence relation. The reader will perhaps have noticed that the cardinal theory contains no mention of this important relation. We saw in Chapter 4 that the relative manyness of two classes depends directly on the type of correspondence which obtains between them. For any two classes R and r', the two classes have equally many terms if the correspondence between them is one-to-one, R has more terms if the correspondence between them is many-to-one, and R has fewer terms if the correspondence between them is one-to-many. Since manyness is not a primitive property of classes but depends on the nature of the correspondence between classes, it would seem more reasonable to predicate the cardinal theory on perceived counterparts of correspondence. There are numerous examples of one-to-one, one-to-many, and many-to-one correspondence in our everyday environments—of which even very young children are aware—that could serve as points of departure. Recall the basketball and football examples from Chapter 4. Also, every monogamously married husband has one wife, and conversely; every child has many toys; several items of equipment belong to each baseball player, and so forth. Since examples of all three types of correspondence are so common in our physical surroundings, it

FIGURE 6. 2

Two Elementary Methods of Eliminating the Relative Manyness Relative Density Confound.

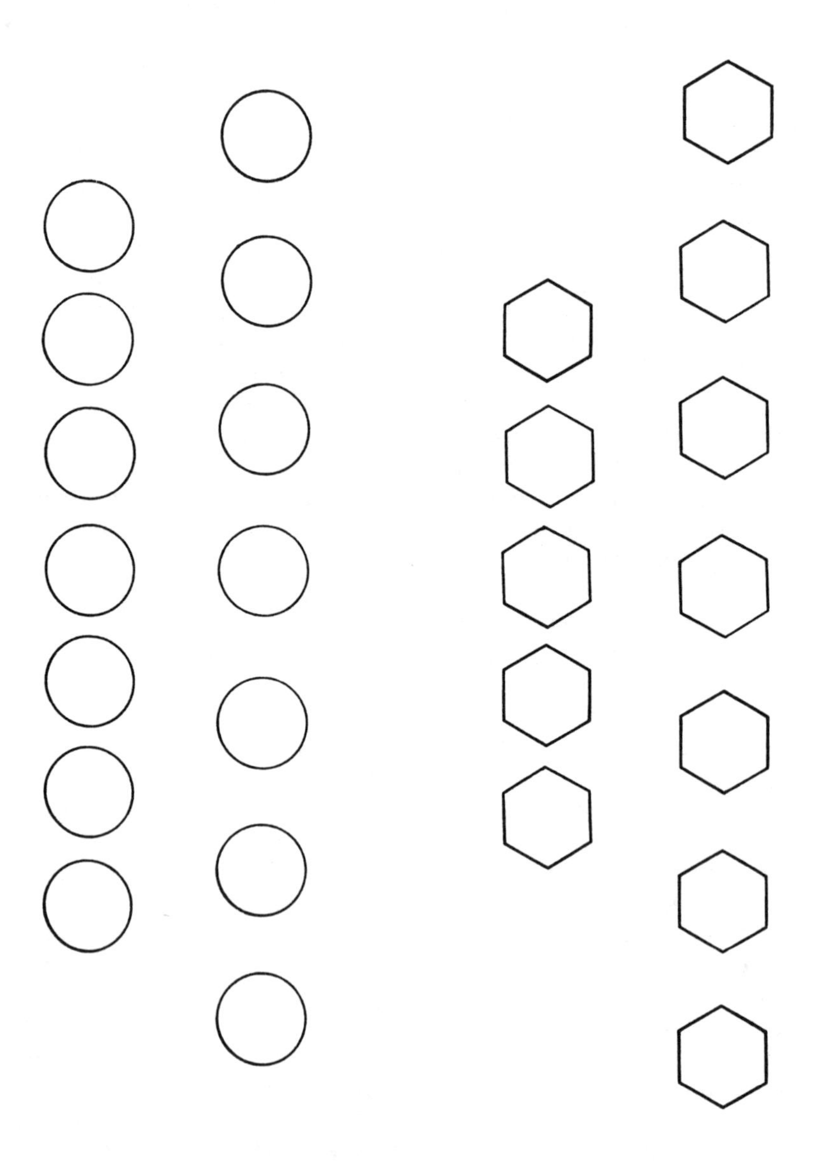

Source: Prepared by the author.

does not appear unreasonable to posit that correspondence, rather than manyness, is the perceptible fact from which cardinal number is derived. Revised in this way, the theory would posit that, first, we are aware of characteristic correspondences obtaining between certain sets of objects and that, second, we gradually become aware that these correspondences imply certain manyness relations between the collections. From this point, the cardinal theory would proceed as above. If the cardinal theory were revised in such a way that manyness was derived from experiences of correspondence, rather than being viewed as a primitive perceptible quality, there would be two important benefits. First, the cardinal theory would be brought more closely in line with the Frege-Russell theory. The Frege-Russell theory, after all, is being accepted as the appropriate logical model of the number concept. In view of the pivotal status of correspondence in this theory, any psychological theory which takes it as a model but yet contains no mention of correspondence must be considered, at best, incomplete. The second benefit which accrues from basing manyness on perceived correspondence is that we could get around, at least for the present, the problem that the operationalization of numerosity has confounded perceived manyness with perceived density.

A third and final difficulty with the cardinal theory is that the concept of numerosity is purely a "difference concept." It is not, strictly speaking, the psychological counterpart of manyness so much as it is the psychological counterpart of dissimilarity. It will be recalled that two questions are posed in the Frege-Russell theory: "What does it mean to say that two classes have different numbers of terms?" and "What does it mean to say that two classes have the same number of terms?" However, the cardinal theory, for some unknown reason, is concerned only with the first of these questions. It provides us with a perceptual hypothesis about how we first come to understand that certain collections contain different numbers of terms. This hypothesis provides a perceptual variable (numerosity) by virtue of which we can decide that two collections of different manyness are instances of, for example, 3 and 5, respectively. But it provides us with no psychological counterpart of similarity: It says nothing whatsoever about how we come to decide that two ostensibly different collections are instances of the same natural class. From the standpoint of the Frege-Russell theory, this is a critical oversight. We have already seen that if this theory is to work, then we must be able to define "same number" and "different number" independently. The definition of the former is especially important because the thesis that a number is anything which is the number of some class turns on this definition. Therefore, even if the first two difficulties with the cardinal theory could be ignored safely, we would still be left with the problem that the theory provides a proposed basis for inferring that two classes have unequally many terms, but it ignores the question of how we infer that two classes have equally many terms.

The preceding problem could be eliminated if correspondence was substituted for numerosity as the basis for manyness judgments. If this were done, we would have a psychological basis for similarity judgments as well as dissimilarity judgments. Whenever we perceive that the correspondence between two collections is one-to-one, as with pitchers and catchers or husbands and wives, we have empirical grounds for inferring "same manyness." Whenever we perceive that the correspondence between two collections is one-to-many or many-to-one, as with mothers and their children or grandchildren and their grandparents, we have empirical grounds for inferring "different manyness." Whether we can or do make the connection between knowledge of correspondence and manyness is, of course, another question, and a question which, among others, the second theory deals with.

A fourth and final difficulty with the cardinal theory is that it contains no direct mention of the ordinal meaning of the number concept. Since the ultimate psychological reference of the natural numbers is cardinal, it is, presumably, safe to conclude that children should acquire an understanding of ordinal number after they acquire cardinal number and after they have learned the basics of arithmetic. Although these seem to be reasonable inferences from the theory, it is nevertheless true that the theory makes no direct mention of ordinal number. It is almost as though proponents of the cardinal theory are unaware that there is an alternative logical model for the number concept.

THE ORDINAL THEORY

The ordinal theory is the most recent of our three theories (Brainerd 1973a, 1973b, 1973c), and is structurally similar to the cardinal theory. Of course, there are some important content differences between the two. The ordinal theory is inspired by the relational approach to number considered in Chapter 3. The theory assumes that number refers to the terms of transitive-asymmetrical relations and that the natural numbers are, in their most basic sense, devices for representing the terms of the progressions which such relations generate. Hence, the psychological origin of number is identified with the psychological origin of ordinal number. Like the cardinal theory, the ordinal theory posits certain prenumerical, perceptual experiences in which the number concept is rooted. In place of numerosity, however, it is proposed that there are examples of the logical notion of progression in the physical environment and that under certain conditions, humans can perceive the transitive-asymmetrical relations which underlie them. Percepts of this sort are called "ordinality." Perceived ordinality, it is argued, is the prenumerical perceptual basis from which the number concept springs. Ordinality may be defined somewhat more concretely as the perception of everyday transitive-asymmetrical relations such as "taller than," "larger than," and "louder than" which underlie progressions found in the real world. Some illustrative progressions are shown in Figure 6.3. All of these progressions are based on the relation "larger than." Unlike the progressions of logic and mathematics, most progressions found in the real world are comprised of relatively few terms—almost always fewer than 10 terms. It is important to note in this regard that a progression must always contain at least three terms. If there are only two terms, then there is no ordering because, while asymmetry is present, transitivity is not. For a given collection of objects to constitute a physical counterpart of the logical notion of progression, there must be three or more objects in the collection.

To summarize the perceptual basis for number, ordinality may be operationally defined as the perception of the ordering relation inherent in some physical progression—where a physical progression always contains at least three terms. The constraints of the environment being what they are, most physical progressions which give rise to perceived ordinality consist of 10 or fewer terms. Finally, it is worth emphasizing that ordinality is exclusively concerned with the perception of the transitive-asymmetrical relations inherent in progressions and not the perception of the individual terms.

It is known that both adults and very young children are capable of perceiving the ordering relations inherent in everyday progressions. To illustrate, consider row A in Figure 6.4. The five dots in row A are arranged from left to right in order of increasing area. Suppose an experimenter and a preschool child (or an adult subject for that matter) are seated across from each other at a table. A card on which row A is printed is placed face-down in the center of the table. The experimenter turns the card face-up and allows the subject to look at it for a very brief interval—say, two seconds. Next, the experimenter introduces a test card on which rows A, B, C are printed and asks, "Which one of these pictures is the set of dots you just saw?" Now, all three pictures are correct because each contains precisely the same five dots as the original card. However, subjects invariably choose the picture in which the five dots are ordered. This suggests that the subjects are perceiving the underlying transitive-asymmetrical relation and not the individual terms in the progression. A further test seems to confirm this suggestion. Suppose that another subject is given a two-second exposure to row A, but is given a test card on which rows B, C, and D are printed. The subject will unhesitatingly choose row D. However, note that the only thing that is the same about the

FIGURE 6.3

Some Empirical Progressions Which Lead to the Visual Perception of Ordinality.

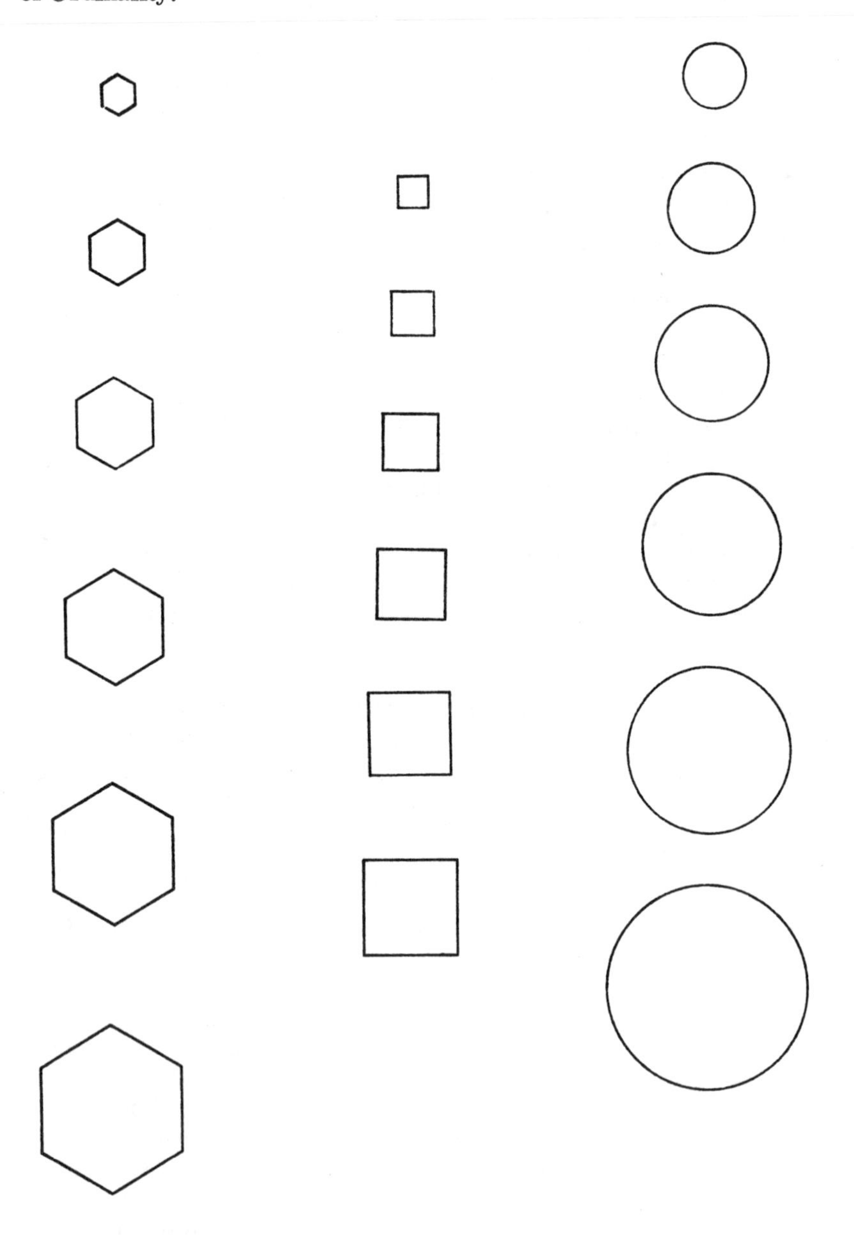

Source: Prepared by the author.

FIGURE 6.4

Four Stimuli Employed to Demonstrate that Children Perceive Transitive-Asymmetrical Relations.

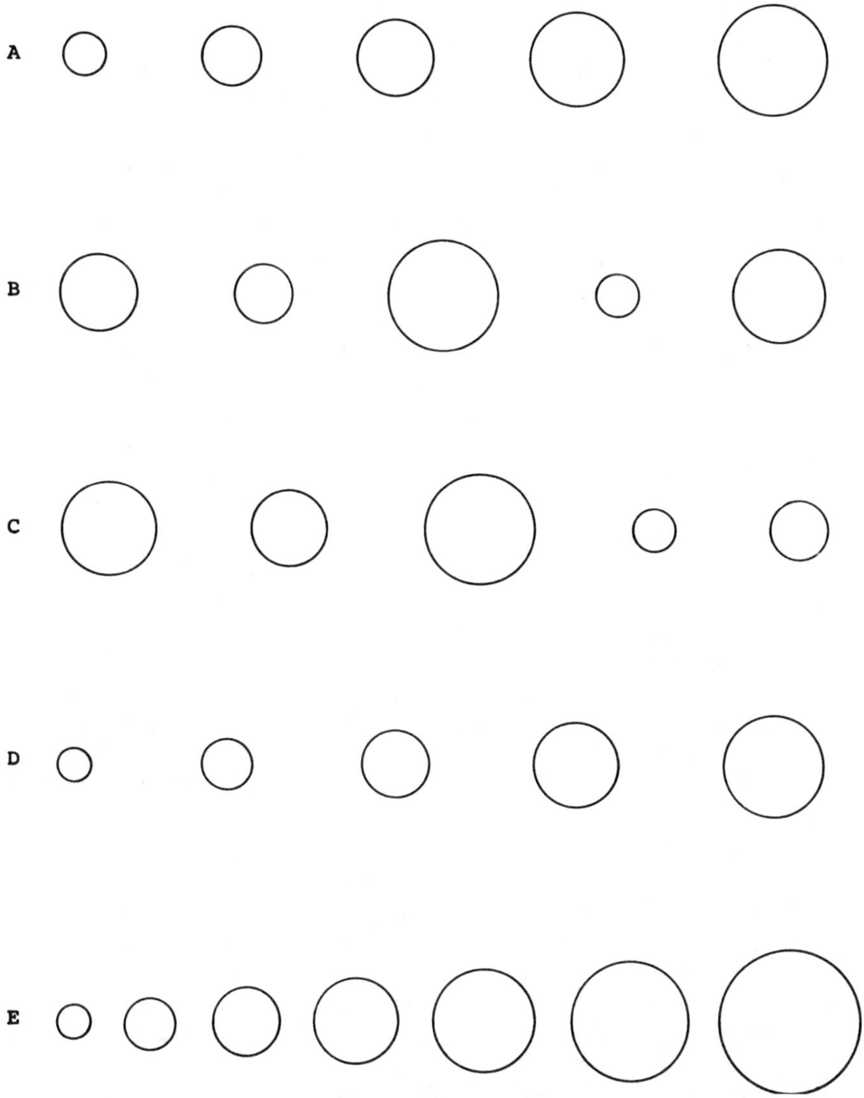

A

B

C

D

E

Source: Prepared by the author.

dots in rows A and D is that they are ordered by area. Note also that no dot in row D is the same size as any dot in row A. In contrast, each dot in rows B and C is identical to one dot in row A. It would seem that the only possible basis for choosing D over B and C is perception of the ordering relation. This conclusion is strengthened by yet a third test. Another subject is given a two second exposure to row A, but is given a test card on which rows B, C, and E are printed. Both the cardinal numbers and the sizes of the individual dots are the same in rows B and C as in row A. In row E, on the other hand, there are more dots than in row A and no dot is the same size as any dot in row A. However, the subject will invariably choose row E.

All this suggests that there is a perceptual version of the logical idea of transitive-asymmetrical relation which we have been calling ordinality. Given three or more concrete objects ordered by some common transitive-asymmetrical relation, even young children can, if the differences between adjacent terms are perceptually salient, perceive the relation itself. It is with this perceptual fact, the ordinal theory maintains, that the psychological concept of number begins. The next step up from perceived ordinality is the psychological counterpart of what was called ordinal number in Part I. The theory posits that gradually, during the preschool years in Western children, the perception of ordinality becomes internalized and takes on the status of a concept. This conceptual descendent of ordinality is called "ordination," and ordination is the psychological counterpart of ordinal number. The presence of ordination in children's thinking is distinguished by the advent of internal ordering skills—that is, ordering skills which, like ordinality, involve everyday transitive-asymmetrical relations but which, unlike ordinality, involve mental representation rather than direct perception of the relations. Ordination is distinguished, first, by the ability to order collections of three or more terms when the inherent ordering relation is not directly perceptible. For example, suppose we present a child with three balls of clay which look exactly alike, as in the illustration at the top of Figure 6. 5. Although the balls look the same, one weighs 50 grams, one weighs 100 grams, and one weighs 150 grams. The child's task is to order the three balls according to weight by comparing the balls in pairs. If the child can solve this problem reliably, then we say that he is capable of ordination—at least with respect to the relation "heavier than." An even more definitive test for ordination involves having the subject deduce the relationship between two of the objects from the relationships between the other objects. We know from logic that the fact that asymmetrical relations such as "heavier than" are always transitive is responsible for the generation of progressions. Indeed, the fact that R_{ac} always follows from R_{ab} and R_{bc} whenever R is asymmetrical is usually called the "minimum ordinal proposition" in logic (Russell 1903). If perceived

FIGURE 6.5

A Simple Method for Assessing Children's Grasp of the Concept of Ordination.

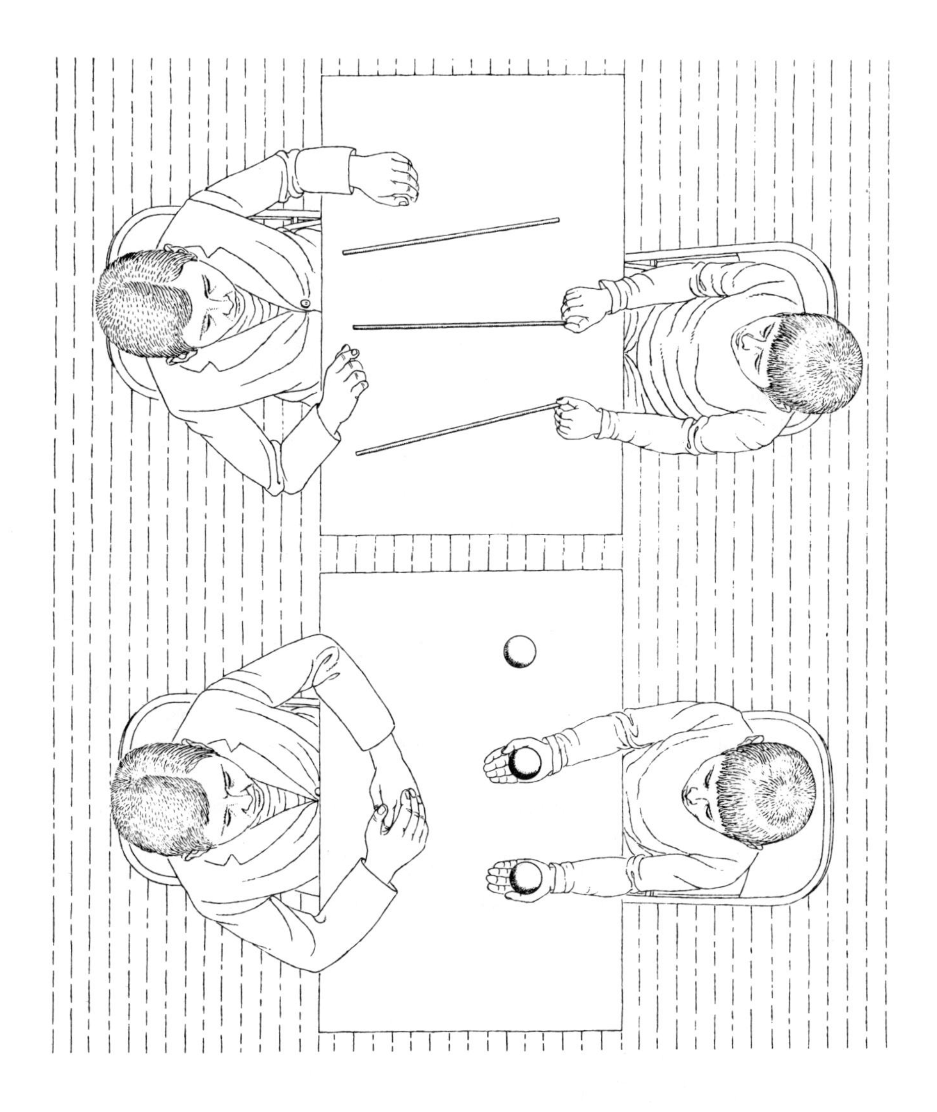

Source: Prepared by the author.

ordinality has truly become an internal concept, it is reasonable to expect a grasp of the fact that, for any three terms of some progression, if a certain asymmetrical relation holds between the first and second terms and between the second and third terms, then it must also hold between the first and third terms. In the case of our clay-ball illustration, we could test the child's understanding of this fact by having the child deduce the relationship between the 50-gram ball and the 150-gram ball after comparing the 50-gram ball to the 100-gram ball and the 100-gram ball to the 150-gram ball. If the child can solve this problem reliably, then it seems safe to conclude that perceived ordinality has become an internalized concept because the relationship between the 50-gram ball and the 150-gram ball was never directly perceived.

Another simple test for ordination is shown at the bottom of Figure 6.5. Our subject is given three sticks which appear to be exactly the same length. Actually, the sticks differ by tiny amounts. The differences can be discovered by carefully comparing the sticks. Can the subject infer the length difference between the first and third sticks by comparing the first stick with the second and the second stick with the third? If this can be done repeatedly, then we say that an understanding of the ordination of length has been achieved.

As might be supposed, the ordinal theory posits that the number concept evolves from the concept of ordination. As perceived ordinality becomes a stable internalized concept, there arises an obvious need for a shorthand symbolism of some sort which may be used to represent the terms in any progression—whether perceived or inferred. In Western cultures, the written numerals, "1," "2," "3," . . . (plus their verbal labels), are available for this prupose. The important point to bear in mind is that whatever specific written or verbal symbols are employed, their original purpose is simply to symbolize the fact of order. At first, these symbols denote only the terms in the small finite progressions found in everyday life. Eventually, as the symbols become less dependent on concrete content, they may be combined to derive much larger progressions than those occurring in the real world. It is with such progressions that arithmetic and the rest of classical mathematics are concerned. Thus, the ordinal theory posits a continuous and gradual transformation of the perception of transitive-asymmetrical relations in the physical environment into the abstract concept of order on which arithmetic is founded. This account, unlike the cardinal theory, is not confronted with the problem of explaining why arithmetic is predicated on a property of the natural numbers other than their original psychological meaning. In the ordinal theory, the original psychological meaning of the number concept and the property of numbers on which arithmetic is founded are the same.

Whereas the ordinal meaning of the number concept is simply ignored in the cardinal theory, the ordinal theory has some explicit things to say about the cardinal meaning of the number concept. In particular, it says something about the developmental relationship between the ordinal and cardinal meanings, and it also says something about the relation of each to arithmetic. The ordinal theory posits a psychological analogy of cardinal number, which is called "cardination." It will be recalled from our earlier discussion of the cardinal theory that there appear to be many everyday examples of one-to-one correspondence, one-to-many correspondence, and many-to-one correspondence of which even very young children are aware. From these everyday correspondences, the ordinal theory proposes, the child eventually works out the connection between correspondence and manyness. Gradually, the connection between correspondence and manyness becomes a stable internal concept. This is the concept of cardination. Generally, the concept of cardination is operationally defined as "the understanding that two classes have equally many or unequally many terms because of the type of correspondence that obtains between them." Moreover, because cardination, like ordination, is a concept rather than a percept, the correspondence between the classes cannot be directly perceptible. The formal similarities between the concepts of ordination and cardination are, it seems, sufficiently obvious to not require further comment.

The ordinal theory makes two important predictions about the concept of cardination. First, the theory stipulates that cardination is not grasped by most children until after they understand ordination. While ordination is acquired during the preschool years by Western children, and it has become a reasonably stable concept by the time most of these children enter elementary school, the same is not true of cardination. Cardination, the theory proposes, is acquired primarily during the elementary school years by Western children, and it does not become a reasonably stable concept until roughly age 9 or 10. The latter range, the reader no doubt will have noted, is considerably later than the age at which these same children are expected to begin learning arithmetic. However, the range for ordination is below the age at which children first are taught arithmetic. These facts lead to a second prediction: Since most children do not grasp cardination until sometime after they begin learning arithmetic, it follows that the cardinal meaning of numbers cannot be very closely connected with the evolution of arithmetic. In particular, it is not generally used to give meaning to written statements such as $1 + 1 = 2$ or verbal expressions such as "three plus two make five." If these statements are to have any initial meaning at all, we must look elsewhere. Since most children do grasp ordination before they are expected to begin learning arithmetic, the ordinal meaning of numbers is available to lend meaning to statements such as the preceding, and the ordinal theory posits that this is, in fact, what happens.

In brief, the ordinal theory may be described as a three-phase account of the growth of the human number concept in which a grasp of the psychological counterpart of ordinal number precedes a grasp of the first elementary statements of arithmetic which, in turn, precedes a grasp of the psychological counterpart of cardinal number. Although these steps are believed to be invariant, the theory does not posit that they are rigid and exclusive. Quite to the contrary, the most reasonable hypothesis is that they overlap considerably. Thus, what the theory actually says is that most children will have made considerable progress with the notion of ordination before they make much progress with arithmetic, and most children will have made considerable progress with arithmetic before they make much progress with cardination. In addition, it is explicitly contended that children's ability to begin learning arithmetic will depend in large measure on their prior understanding of ordination but will not necessarily be closely connected with their grasp of cardination.

PIAGET'S CARDINAL-ORDINAL THEORY

Piaget's theory is similar to the ordinal theory in that it deals with both the ordinal and cardinal meanings of the number concept. It differs from the ordinal theory in three major respects. First, the theory focuses on the growth of ordination and cardination as concepts and does not make any explicit proposals about the perceptual basis of either. No attempt is made either to ground ordination in the perception of ordering relations or to ground cardination in the perception of correspondence or manyness. Instead, the emergence of both concepts is explained in terms of the acquisition of eight hypothetical cognitive structures, called "groupements," which are presumed to appear during the late preschool and early elementary school years (Piaget 1942, 1949). In lieu of a more thorough analysis of these cognitive structures, which is far beyond the scope of this inquiry, the relationship between the structures, on the one hand, and ordination-cardination, on the other, may be said to be roughly the same as the relationship between a computer program and its outputs. The structures, like the program, make certain conceptual operations possible. Ordination and cardination happen to be two of these. A second distinguishing feature of Piaget's theory is that it contains few systematic references to arithmetic. Although, as we shall see below, predictions about the conceptual precursors of arithmetic follow from the theory, they are nowhere explicit in the theory. Also, unlike the cardinal and ordinal theories, when Piaget uses the phrase "natural number," he does not mean arithmetic, but usually means the concept of cardinal invariance (Beth and Piaget Ch. 11, sec. 51), which was discussed at some length in Chapter 2. He renames this concept "number conservation." The third distinguishing feature of the theory concerns the role Piaget assigns to the concept of cardination. Whereas the ordinal theory postulates that ordination is more fundamental to the number concept than cardination, Piaget believes that the natural numbers are simultaneously ordinal and cardinal.

Three general predictions about the growth of children's numerical ideas follow from Piaget's theory. Each of these predictions have been discussed elsewhere in the psychological literature on concept development (for example, Brainerd 1973a, Flavell 1970). First, there should be no developmental lag between the appearance of ordination in children's thinking and the appearance of cardination. Since the number concept is equal parts ordinal and cardinal, ordination and cardination must emerge synchronously rather than asynchronously. This is not to say that individual children always acquire these two concepts synchronously. However, the order in which they are acquired is always idiosyncratic to a particular child and, therefore, in large samples of children, no single order should dominate. The age range which Piaget proposes for the development of ordination and cardination in Western children is the late preschool and kindergarten years. Thus, we may expect that most first and second graders, and many kindergarteners, will evidence reasonably good understanding of both concepts. A second prediction is that ordination and cardination both play central roles in the development of arithmetic. Third, the average child must understand both ordination and cardination before making much progress with arithmetic.

Although there is general agreement that the first two predictions follow from Piaget's theory, some investigators are uncertain about whether or not the third prediction follows. For example, John H. Flavell, one of North America's most respected Piaget analysts, once confessed that "It is unclear (at least to the writer) whether Piaget's theoretical analysis of the number concept as a synthesis of classificatory and relational operations actually ought to imply that the standard Piagetian tests for these operations . . . should be negotiated earlier than, or in strict synchrony with, more direct tests of the concept itself [arithmetic computation] . . ." (Flavell 1970: 1003).

The uncertainty about the predicted relationship between ordination-cardination and arithmetic is, it seems, almost entirely a consequence of two curious features of Piaget's number writings. First, as noted earlier, Piaget's theory does not deal directly with children's understanding of arithmetic. Piaget employs number conservation as his "minimum criterion for the acquisition of number" (Beth and Piaget 1966: 259), whereas, in our first two theories and among psychological investigators generally, arithmetic is the index of natural number. The use of number conservation as the criterion of natural number entails that most of Piaget's statements about the developmental relationship between ordination-cardination and number (Piaget 1952: 147-57) actually are concerned with the developmental relationship between ordination-cardination and number conservation. As we shall see, the predictions about ordination-cardination vis-à-vis number conservation are not the same as ordination-cardination vis-à-vis arithmetic. Second, Piaget's principal treatise on the number is extraordinarily obscure when it comes to making hard predictions about the order in which children acquire numerical ideas (1952). There are even some contradictory predictions in this book. Fortunately, Piaget has made his predictions considerably less equivocal in subsequent and, alas, lesser-known works (Beth and Piaget 1966, Piaget 1970).

Despite the uncertainties, it appears fair to say that Piaget's theory does and must predict that ordination and cardination both precede arithmetic. There are at least three grounds for this conclusion. The first one is somewhat elliptical. In his most recent number writings, particularly *Mathematical Epistemology and Psychology*, Piaget has averred in what, for him, are clear terms that (1) ordination, cardination, and number conservation all emerge in tight synchrony and that (2) children must grasp number conservation before they can make progress with arithmetic (on the second point, see also Piaget 1952, Pt. 3). Now, if ordination and cardination are acquired at the same time as number conservation and number conservation precedes arithmetic, it follows that ordination and cardination also precede arithmetic. Second, either the developmental priority of ordination-cardination over arithmetic must follow from Piaget's theory or the theory simply is not an acceptable psychological explanation of the origins of the number concept. What we require in a psychological theory, it will be recalled, are hypotheses about what concepts are necessary prerequisities for being able to do arithmetic. Any such hypothesis leads automatically to the prediction of a developmental lag. The final reason is the most obvious of the three. It will be recalled that Piaget usually specifies the late preschool and kindergarten years as the age range during which Western children acquire ordination and cardination. We know that these same children will not begin evidencing much arithmetic competence until the first and second grades. Hence, if it is assumed that Piaget's proposed age range is correct, then ordination-cardination must precede arithmetic.

To conclude our examination of Piaget's theory, it would be helpful to summarize Piaget's description of the development of cardination and ordination. He proposes that the development of both concepts consists of the same three stages. The evaluation of the three cardination stages centers on the measurement of certain classification skills.

During Stage "one" Piaget claims that children are incapable of exhaustively classifying a collection of concrete objects. For example, suppose the child is given a collection consisting of five red poker chips, five blue poker chips, five green poker chips, and is asked to "put the things together which go together." An adult would immediately sort the chips into three piles of five each, but the Stage 1 child arranges the chips into pictures of common objects which Piaget calls "figural collections." During Stage II, Piaget claims that children can solve the sorting problem, but they do not yet grasp the principle of class inclusion. The class-inclusion principle is assessed by presenting the child with two collections of unequal cardinality (such as 6 toy cows and 10 toy horses) and asking about the relative cardinality of the largest collection and the superordinate class ("Are there more horses or more animals?"). During Stage 3, Piaget claims that children are capable of both exhaustive classification and class inclusion. The principle of class inclusion, because it is associated with Stage 3, is Piaget's sine qua non of cardination.

Piaget's evaluation of the three stages of ordination centers on skills such as those mentioned in conjunction with the ordinal theory. Suppose the child is given a collection of three or more objects between which some transitive-asymmetrical relation obtains and is asked to order them. Moreover, assume that the relation is not immediately perceptible but, rather, must be discovered by comparing the objects in pairs. During Stage 1, Piaget claims that children are incapable of arranging the terms of such collections in order. He reports that they can determine the pairwise asymmetries but that the overall order escapes them. Piaget's Stage 2 children, on the other hand, can arrange all the terms in order. However, he claims that they are not yet aware of the fact that an asymmetrical relation is always transitive. For any three objects between which a transitive-asymmetrical relation obtains that is not immediately perceptible, Piaget's Stage 2 children cannot reliably infer the relation between the first and third terms once they know the relation between the first and second terms and the second and third terms. However, Piaget's Stage 3 children solve this problem. As was the case for the ordinal theory, therefore, Piaget's ultimate criterion for the concept of ordination is the minimum ordinal proposition of logic.

In brief, Piaget claims that the development of ordination and cardination are characterized by the same three global stages. From this fact and from the fact that the three stages span roughly the same age range, Piaget concludes that ordination and cardination are acquired synchronously.

PART II

THE DEVELOPMENT OF NUMBER

7

A PRÉCIS OF PAST RESEARCH AND SOME OPERATIONAL DEFINITIONS

We now have three contrasting accounts of the growth of children's number concepts which correspond in a more than coincidental way to the logical doctrines reviewed in chapters 3, 4, and 5. The oldest and most widely acknowledged of the three posits that the human number concept, as indexed by arithmetic behavior, arises from a prior understanding of the concept of cardination. Cardination, in turn, arises from perceived numerousness, and perceived numerousness presumably is part of our natively given perceptual apparatus. No explicit role is assigned to ordination. The most recent and least known of our theories posits that number arises from prior understanding of the concept of ordination and that cardination is not very closely connected with number. Ordination presumably is rooted in perceived ordinality, which, in all probability, is a natively given percept. Piaget takes yet a third approach in which the number concept arises from a prior understanding of both ordination and cardination. Unlike the ordinal and cardinal theories, Piaget does not explain the origins of ordination and cardination in terms of their perceptual precursors. He resorts, instead, to mentalistic entities called "cognitive structures" to explain ordination and cardination (Piaget 1942).

It is now time to see what research on the development of children's number concepts tells us about the predictions of the three contrasting theories. In this brief chapter and the three which follow it, we shall examine data that, it is hoped, will permit us to decide which of the theories is the most promising. In this particular chapter, we consider two preliminary matters. First, we shall review studies of number development, conducted primarily during the past two decades, that are relevant to the predictions of our theories. We shall see that although the development of children's number concepts has been a popular research topic during recent years, data bearing directly on the predictions we are interested in are extremely thin and inconclusive. Second, we consider how best to go about measuring the concepts of ordination and cardination. We shall see, in the case of cardination, that there are some bothersome extraneous variables which must be controlled. In Chapter 8, the presentation of evidence continues with three large-scale normative studies involving several hundred children. The data of these studies are concerned with the developmental predictions of our three theories. In Chapter 9, two experimental studies are reported whose data are concerned with the functional predictions of the theories. The presentation of evidence concludes in Chapter 10 with some data concerned with the development of cardination during the elementary school years.

OVERVIEW OF PAST RESEARCH

We turn now to studies of numerical development whose findings apply, for the most part indirectly, to the problem of deciding which of our three theories is most nearly correct. It should be stressed that it is in no way the purpose of this section to review the existing literature on number development comprehensively. In the first place, reasonably exhaustive reviews of the number development literature already are available (Flavell 1970, Gelman 1972). Further, only a segment of this literature—and a very small segment at that—is relevant to the problem at hand. As a consequence of the influence that Piaget's original work has had, most studies of number development conducted in recent years have dealt exclusively with cardinal number. The dominant theme in these studies has been to determine the perceptual cues which children employ as bases for inferring that two classes have equally many or unequally many terms (Lawson, Baron, and Siegel 1974; Pufall and Shaw 1972; Smither, Smiley, and Rees 1974). Most of these studies have also been concerned with Piaget's number conservation concept. Although these studies tell us a great deal about how the cardination concept develops, they do not tell us anything about developmental lags between ordination, cardination, and arithmetic, nor do they tell us anything about the functional dependence of arithmetic on ordination and cardination. For our purposes, therefore, these studies are not very informative. We are interested in findings that tell us something about the developmental relationships between ordination and cardination, ordination and arithmetic, and cardination and arithmetic. We also are interested in the functional relationships between these variables.

Of the many studies of children's numerical concepts that are now available, only the investigations of Beard (1963), Dodwell (1960, 1961, 1962), Hood (1962), Siegel (1971a, 1971b, 1974), Wang, Resnick, and Boozer (1971), Beilin and Gillman (1967), and Piaget's original investigations (Piaget and Szeminska 1941), appear to provide information relevant to the present inquiry. Piaget's classic research, though by far the most ambitious of the aforementioned studies, provides data which are primarily concerned with the developmental relationship between the concepts of ordination and cardination. The Child's Conception of Number is divided into three parts. Part II, which consists of four chapters, deals with the growth of ordination and cardination. In the first two chapters, Piaget introduces some tests of cardination (correspondence between glasses and bottles, correspondence between flowers and vases) and reports some typical responses to the tests. At the end of the second chapter is presented the three-stage model of cardination development outlined in Chapter 6. In the third and fourth chapters of Part II, Piaget introduces some tests of ordination (seriation, serial correspondence) and reports some illustrative responses. Near the end of the fourth chapter, he presents the three-stage model of ordination development considered in Chapter 6. Also near the end of the fourth chapter, he discusses the developmental relationship between ordination and cardination. Piaget concludes that, generally speaking, there is no developmental lag between the two concepts.

It must be admitted that the actual data on which Piaget's conclusion about the order of emergence of ordination and cardination are based are extremely problematical. In fact, they are so questionable that his conclusion seems premature at best. To begin with, he makes absolutely no direct statistical comparisons of his subjects' performances on the ordination and cardination tests. The subjects were administered either the ordination or cardination tests but never both. Thus, no evidence is reported about the relative performance of the same subjects on the different tests. Instead, there are only data about the absolute performance of different subjects on different tests. Such data preclude direct within-subject statistical comparisons which, in cross-sectional research, are the standard method of investigating possible developmental lags between two or more concepts. As it turns out, Piaget's only basis for concluding that ordination and cardination emerge synchronously is that the same three global stages were observed for both concepts and the stages spanned roughly the same age range. (Virtually all of Piaget's subjects were between four and six years old). Piaget recognized that such evidence is less than sufficient to establish the synchrony of ordination and cardination. However, he justified his choice of evidence on the curious ground that doing things correctly would have been too tedious:

> The same processes and the same levels are to be found in the development of both ordination and cardination. But obviously any attempt to express the situation in statistical form and to apply correlation formulae to these tests would involve us in questions for which we must confess we have

> little interest. . . . The calculation of the correlation between the levels of cardination and ordination, without the accompaniment of an extremely thorough qualitative analysis, could therefore give only misleading results, unless our experiments were transformed into 'tests' in which statistical precision could no doubt easily be obtained, but at the cost of no longer knowing exactly what was being measured (Piaget 1952: 149).

It is something of a testimony to the primitive state of research on human cognitive development that this convoluted and somewhat self-serving rationale has been accepted with minimal protest to date. At the very least, it would seem that Piaget might have told us the average ages of the subjects he deemed to be functioning at each of the ordination stages and each of the cardination stages. He did not, however. One is therefore forced to conclude that Piaget's data tell us little, if anything, about the developmental relationship between ordination and cardination.

Things might not be any clearer if Piaget had made the appropriate within-subject comparisons. Without going into the specifics of his ordination and cardination tests, it can simply be noted that there was no attempt to construct tasks that are precise embodiments of the definitions of these concepts given in the preceding chapter. The relation between Piaget's tasks and the formal definitions of ordination and cardination is one of very loose analogy rather than strict operational definition. Presumably, this is because the aims of the research were exploratory. The inevitable implication, however, is that, in addition to telling us very little about the developmental relationship between ordination and cardination, it is not clear as to precisely what Piaget's data tell us about the development of ordination and cardination separately.

The other studies mentioned above do not suffer from the elementary psychometric flaws which plague Piaget's research. All the tests were given to all the subjects in these studies and, consequently, the appropriate within-subject comparisons could be made. Unfortunately, these studies do tend to share the operational-definition problem of Piaget's studies. For the most part, the investigators chose not to construct new and more precise tests of ordination and cardination. They almost always chose simply to adopt Piaget's original tests. The first of Dodwell's three studies (1960) dealt with children's performance on conservation tests, two of Piaget's cardination measures (provoked and unprovoked correspondence), and two of Piaget's ordination measures (seriation and ordinal-cardinal correspondence). Although the data certainly were available to Dodwell, for some reason no direct evidence on the possible developmental lag between ordination and cardination were reported. Dodwell's second study (1961) dealt with the amount of statistical covariation between the Piagetian tasks administered in the first study and a test of arithmetic comprehension. Children's overall performance on the Piagetian tasks correlated moderately well with their arithmetic comprehension. Importantly, from our standpoint, no attempt was made to determine the precise individual correlations of the conservation, cardination, and ordination tests with arithmetic. Thus, it is possible that the moderate overall correlation could be masking a zero correlation between one of these three tests and arithmetic. The final Dodwell study (1962) was concerned with the amount of statistical covariation between Piaget's sine qua non of cardination, class inclusion, and arithmetic proficiency. All of the correlations observed between the two variables were either low-positive or not significantly different from zero. This led Dodwell to conclude that the relationship between cardination, at least as Piaget prefers to measure it, and arithmetic proficiency is not as pronounced as the Piaget theory leads us to expect.

In the Hood investigation (1962), the relationship between children's arithmetic proficiency and their grasp of ordination, cardination, and some of Piaget's conservation problems was studied. After administering tests of these variables to elementary school children, Hood graded the subjects' performance according to certain previously established levels of proficiency. From lowest to highest, there were five levels in all. Once the data had been graded in this manner, a developmental lag was noted between arithmetic skill, on the one hand, and ordination-cardination-conservation, on the other. All of the children classified at the lowest level of arithmetic competence were classified at a higher level of competence on the remaining tasks, some of these children were classified at the highest level on the remaining tasks. Similarly, none of the children who had attained the highest level of performance on ordination-cardination-conservation had attained the highest level of arithmetic performance. Some of these children were still functioning at the lowest of the five levels of arithmetic. These findings are important because they seem to suggest that either ordination or cardination, or both, may appear in children's thinking before they become very proficient with arithmetic. Unfortunately, certain percularities of Hood's data make this suggestion no more than a tentative one. Although Hood graded the children's arithmetic performance according to five predetermined levels of skill, performance on the ordination and cardination tasks was not individually graded using the same five levels. Instead, Hood used a combined grading system in which the children were assigned to the five levels of "concept" proficiency on the basis of their composite performance on ordination, cardination, and conservation. Therefore, arithmetic competence was not separately compared with ordination, cardination, and conservation. Thus, about all we can say from Hood's data is that children seem to understand some of these latter concepts before they become proficient with arithmetic. There is no way to ascertain whether all of these concepts precede arithmetic or whether the observed developmental lag is restricted to one or two of them.

Beard's study (1963), like Hood's, dealt with the relationship between children's arithmetic proficiency and Piaget's number-concept tasks. Ordination, cardination, and conservation were among the tests in Beard's number-concept battery. Unfortunately, Beard, like Hood, did not analyze his findings in a manner that is very helpful to us. His study was primarily a correlational one that attempted to examine the general statistical relation between arithmetic performance and the overall battery of number-concept tasks. There was no explicit interest in determining whether or not there were any developmental lags between the various concepts. Beard reported that arithmetic performance tended to correlate in a moderate positive manner with performance on the number-concept tasks. However, the correlations between arithmetic skill and performance on the individual tests in the number concept battery—especially the ordination and cardination tests—were not reported.

The studies summarized up to this point appear to suggest, however tentatively, that there may be a positive statistical relationship between children's arithmetic competence and their grasp of ordination or cardination, or both. However, we still are completely in the dark about whether there is a developmental relationship between these variables. Of the various investigations cited at the outset of this brief review, only the studies by Siegel and by Wang, Resnick, and Boozer were explicitly designed to examine the order in which children acquire numerical concepts. Siegel's first two studies (1971a, 1971b) investigated the order in which preschool and elementary school children acquire seven concepts: (1) continuous magnitude, (2) discrete magnitude, (3) equivalence, (4) conservation, (5) ordinal position, (6) seriation, (7) addition.

For our purposes, the most important of the concepts Siegel studied are the last three. To measure each of them, she employed a programmed-learning apparatus similar to the ones that are sometimes used in elementary school classrooms. The child sits in front of an apparatus which consists of a large viewing screen and several buttons that the child can press. The apparatus presents various stimuli on the viewing screen and the child responds to the stimuli by pressing buttons. Whenever a correct response is made, the apparatus automatically provides a reward of some sort (a toy or a candy). On each trial of Siegel's ordinal-position task, the programmed-learning apparatus presented a series of two, three, or four pictures on the screen. The child was instructed to press the button under the second picture and received a reward for a correct response. Thus, on the ordinal-position task, the child was required to respond to a particular position in several finite progressions of terms. On each trial of Siegel's seriation task, the child was presented with a series of three pictures, each containing between one and nine dots. The child was instructed to pick the middle-sized collection of dots, and received a reward if correct. On each trial of Siegel's addition task, a series of five pictures was presented on the screen, each picture containing between one and nine dots. The experimenter arbitrarily designated one of the collections a "sample" and asked the child to determine which of the remaining four collections had two dots more than the sample. Again, all correct responses were rewarded.

Interestingly from the standpoint of the present inquiry, Siegel observed a developmental sequence in the emergence of ordinal position, seriation, and addition. Her subjects understood ordinal position before they understood either seriation or addition. Kindergarten children, for example, required an average of only 16 trials to learn that they must always respond to "second," but they required roughly 57 trials to learn that they must always respond to "middle-sized." The same kindergarteners who required 57 trials to learn "middle-sized" and 16 trials to learn "second" required 73 trials to learn that they must always respond to "greater than by 2." For our purposes, the crucial fact about these findings is that they seem to show that two ordination-like concepts are understood before an arithmetic-like concept. These findings are, of course, consistent with the ordinal view of number development discussed earlier.

Siegel's third experiment (1974) was designed to test the ordinal theory's prediction that children grasp ordination before cardination. Her subjects were 91 children between three-and-a-half and five years old. Siegel did not use a programmed learning apparatus in this experiment. Instead, an experimenter simply showed each child seven different series of 5" x 7" cards. Correct responses were rewarded by giving the subjects coins which could be exchanged for toys at the conclusion of the experiment. Siegel's ordination test consisted of a series of three problems on which the subjects were required to learn a magnitude discrimination. On problem 1, the subjects were shown a series of cards like the one at the top of Figure 7. 1. Each card contained two sets of dots. The number of dots in each set varied, depending on the card, between 2 and 9. The set which contained more dots was always longer than the set which contained fewer dots. Half the subjects were taught to select the less numerous set and half were taught to select the more numerous set. The second ordination problem was exactly the same as the first, except for the cards that the subjects were shown. On this problem, cards like the one in the center of Figure 7. 1 were used. Although the number of dots in the two sets

FIGURE 7.1

Stimuli Employed to Study the Developmental Relationship Between Ordination and Cardination.

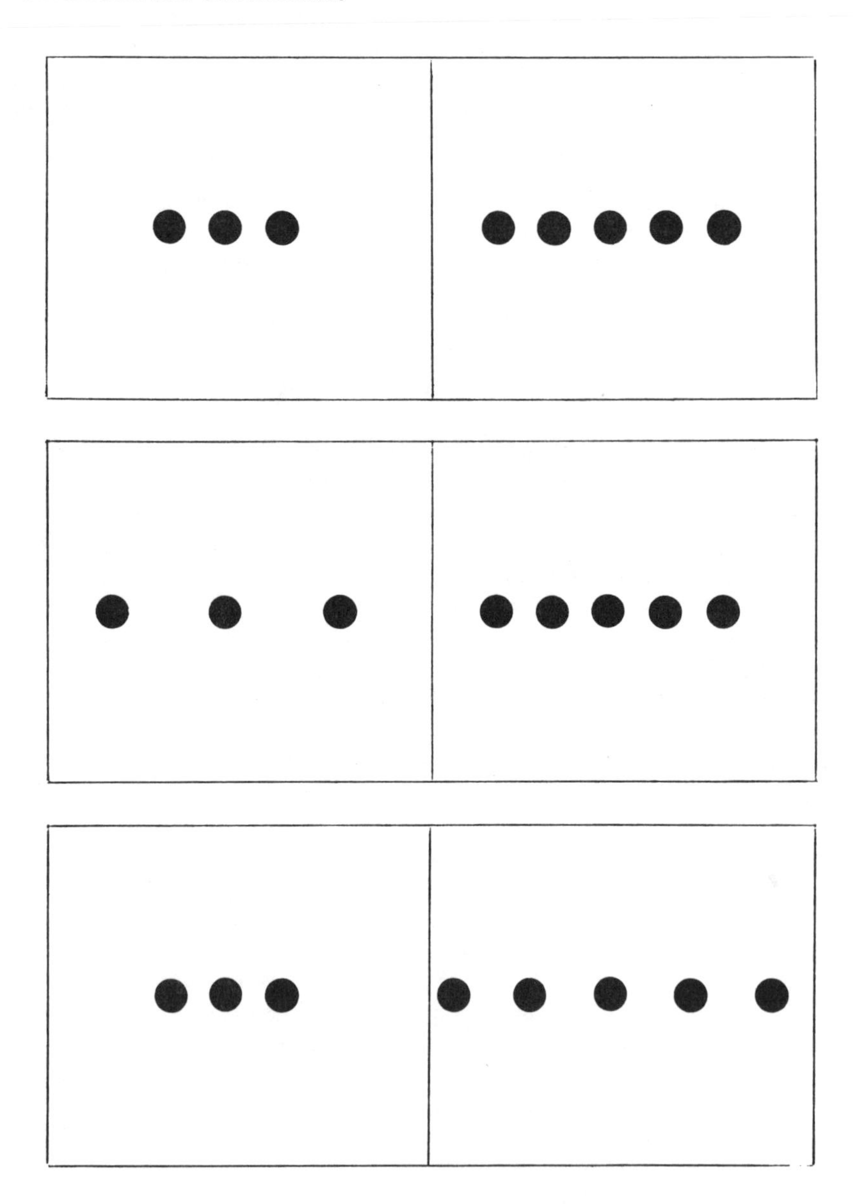

Source: Constructed by the author.

again varied between 2 and 9, the two sets were always the same length and, consequently, the dots in the more numerous set were more densely packed than the dots in the less numerous set. The third ordination problem also was the same as the first one, except for the cards that were used. The subjects were again shown cards with two sets containing between 2 and 9 dots. But in this series the cards resembled the one at the bottom of Figure 7.1. That is, the less numerous set was always shorter and denser than the more numerous set.

Siegel's cardination test consisted of four problems on which the subjects had to learn to identify sets with the same cardinal number. On each problem, the subjects were shown a series of cards resembling those in Figure 7.1. However, each of the cards used on the cardination problems contained three sets of dots rather than two. The experimenter designated one of the sets on each card as a "sample," and asked the child to determine which of the remaining two sets contained the same number of dots as the sample. The four cardination problems were the same, except for the cards that the subjects were shown. On problem 1, the densities of all three sets were the same. On problem 2, the sample and the correct alternative set were the same length as the incorrect alternative set. On problem 3, the sample was not the same length as either alternative set. On problem 4, the sample was the same length as the incorrect alternative set but not the same length as the correct alternative set.

The interesting finding of Siegel's third experiment is that children solved the ordination problems more quickly than the cardination problems. On the average, the children required 3 trials to solve the first ordination problem, 16 to solve the second, and 33 to solve the third. But they required an average of 15 trials to solve the first cardination problem, 14 trials to solve the second, 20 to solve the third, and 27 to solve the fourth. Thus, on the average, the cardination problems were about 12 percent more difficult than the ordination problems. In line with the ordinal theory, Siegel concluded that "quantitative concepts which rely on the understanding of transitive asymmetrical relationships . . . precede the development of cardinal concepts of number" (1974: 911).

The Wang, Resnick, and Boozer study (1971), like Siegel's first two investigations, is a rather complex piece of research concerned with the developmental order in which children acquire a broad spectrum of numerical ideas (16 specific numerical concepts in all). Most of specific concepts studied fall outside the scope of our inquiry. However, the findings on some of these concepts are relevant. The authors presented kindergarten children with a series of tests in which they were required to determine whether two collections contained equally many or unequally many terms. The collections to be compared always contained 10 or fewer terms. The children were asked to judge relative manyness according to two different methods—counting, which is an ordinal procedure, and term-by-term correspondence, the primary cardinal operation. Interestingly, the subjects were able to determine the relative manyness of collections quite easily via the first method. However, they found the second method more difficult to use. From the perspective of the ordinal theory, this finding might be interpreted as suggesting a developmental lag between ordination and cardination.

Finally, we consider Beilin and Gillman's (1967) research. Beilin and Gillman conducted a learning experiment in which children were taught to respond to ordinal and cardinal numbers. The subjects were 105 first graders. The children were divided into two groups—an ordinal number learning condition and a cardinal number learning condition. The reversal shift procedure popularized by Kendler and Kendler (1962) was used to train each child on either ordinal or cardinal number. A series of 3" x 3" cards like the ones shown in Figure 7. 2 was used in both learning conditions. A figure consisting of several rectangles was located in the center of each card. Depending on the card, the figure in the center consisted of four, five, or six rectangles. One of the rectangles in each figure was colored red. On half of the cards, the red rectangle was the second one from the left. On the other half, the red rectangle was the third one from the left.

A three-phase procedure was used in the ordinal-number condition. Each of the first two phases consisted of a series of training trials during which the subjects were trained to respond to position. During the first phase, each child was trained to respond to one and only one of the two ordinal positions of the red rectangle. For example, consider some given child who was trained to respond to the second position from the left. On the first trial, the child might see a pair of cards like those at the top of Figure 7. 2 and be asked to choose one of the cards. The choice of the card on the right is rewarded with a marble; the choice of the card on the left is not rewarded. On the second trial, a pair of cards like those at the bottom of Figure 7. 2 is presented, and the choice of the card on the left is rewarded. During the second phase, each child was trained to respond to the reverse of whatever ordinal position was learned during the first phase. Suppose that a given child was trained to respond to the second position from the left during the first phase. On the first trial of the second phase, a pair of cards like those at the top of Figure 7. 2 was employed. This time, however, choosing the card on the left, rather than the card on the right, is rewarded. Similarly, when seeing a pair of cards like those at the bottom of Figure 7. 2, the child is rewarded for choosing the card on the right rather than the card on the left.

The third phase of Beilin and Gillman's ordinal-number procedure was a test designed to measure how well the children had learned to

FIGURE 7.2

Stimuli Used to Study Ordinal-Number Learning and Cardinal-Number Learning.

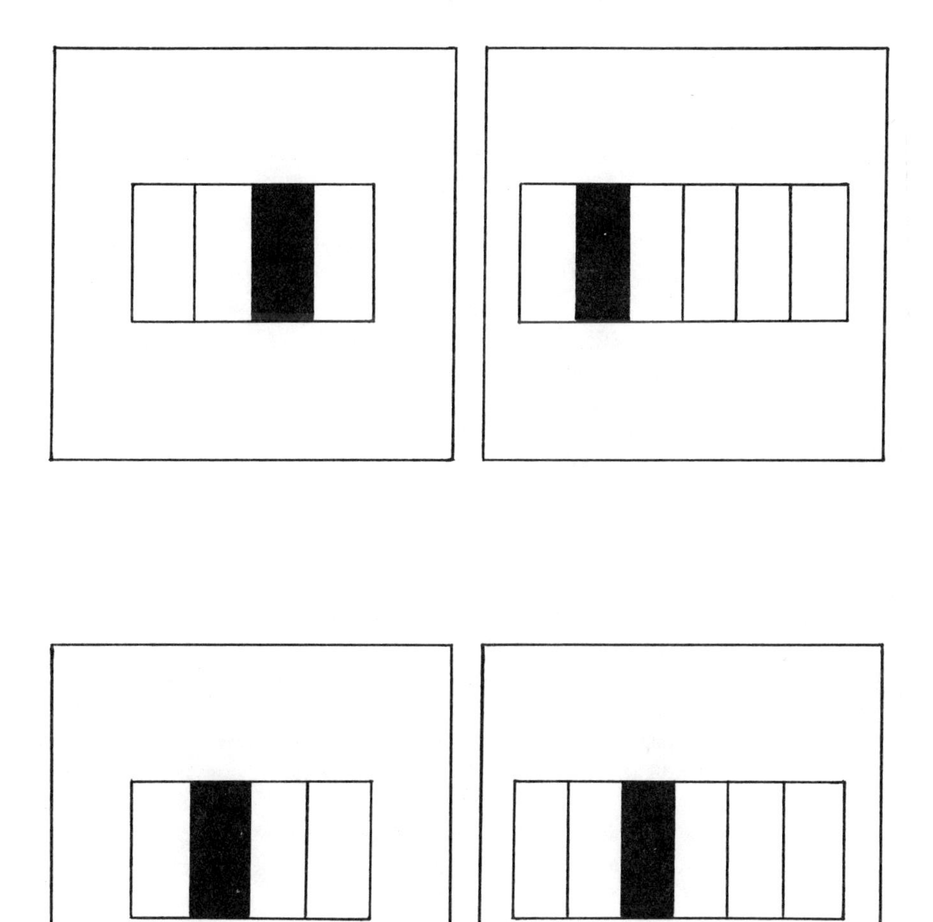

Source: Constructed by author.

respond to position during the first two phases. The strength of the children's positional responding during phase three was Beilin and Gillman's estimate of ordinal-number learning. A similar three-phase procedure was used in the cardinal-number condition. During the first phase, the children were trained to respond to a particular manyness. For example, consider some child who was trained to respond to the cardinal number "4" during phase one. On the first trial, the child sees a pair of cards like those at the top of Figure 7.2 and is asked to choose one of them. The choice of the card on the left is rewarded. The same is true of the pair of cards at the bottom of Figure 7.2: the left-hand card receives a marble. During the second phase, each child was trained to respond to the reverse of the cardinal number the choice of which was rewarded during phase one. Suppose that some given child was rewarded for choosing cards with four rectangles and, thus, on the first trial of phase two, sees the pairs of cards at the top of Figure 7.2. Now, a reward is given for choosing the card on the right. Similarly, given the pair of cards at the bottom of Figure 7.2, it is the right-hand choice which obtains a marble. The third phase of the cardinal-number procedure, like the third phase of the ordinal-number procedure, was a test designed to measure how well the children had learned to respond to manyness during the first two phases. The strength of children's tendency to respond to manyness during phase three was Beilin and Gillman's estimate of cardinal-number learning.

Generally speaking, first graders found both forms of learning to be difficult. In the ordinal-number condition, 67 percent of the subjects failed to show clear evidence of learning. That is, during the phase-three test, two-thirds of the ordinally trained subjects did not yet understand that their task was to respond to the position of the red rectangle. In the cardinal-number condition, 88 percent of the subjects failed to understand that their task was to respond to how many rectangles appeared on the cards. Although both forms of learning were difficult, there was evidence that ordinal-number learning was easier. During phase three, the number of ordinally trained subjects who always responded to position was three times as large as the number of cardinally-trained subjects who always responded to manyness. This finding is very interesting because it is exactly what the ordinal theory would predict. Unfortunately, however, Beilin and Gillman conducted a second experiment and they were unable to replicate their findings. This time the number of ordinally trained subjects who always responded to position during phase three was the same as the number of cardinally-trained subjects who always responded to manyness. Thus, Beilin and Gillman's research provides data which is only suggestive.

It is obvious from the research reviewed in this section that the available evidence on ordination, cardination, and arithmetic is sketchy and not very conclusive. Insofar as the developmental relationship between ordination and cardination is concerned, there is very little solid evidence. With the exception of Siegel and Beilin and Gillman, investigators seem to have accepted Piaget's hypothesis that ordination and cardination develop synchronously as a foregone conclusion not worthy of further examination. Unfortunately, Siegel's and Beilin and Gillman's findings are inconclusive. We have slightly more evidence about the order of acquisition of ordination-cardination and arithmetic. Hood's data (1962) are vague, while Siegel's data (1971a, 1971b, 1974) are stronger, in support of the notion that ordination may be acquired before arithmetic. However, the value of Hood's data are greatly diminished by the aforementioned peculiarities in the data analysis, and the value of Siegel's otherwise excellent studies is limited by the fact that she studied only a very narrow aspect of arithmetic proficiency (adding 2). To verify the relationship between ordination and arithmetic observed by Siegel, we shall require a much broader and more comprehensive method of measuring arithmetic skill. Finally, concerning the developmental relationship between cardination and arithmetic, there are conflicting findings. The data of Hood's study suggest the same relationship between cardination and arithmetic as between ordination and arithmetic. However, Dodwell's evidence (1960, 1961, 1962) indicates that arithmetic skill and cardination may develop independently.

As if the existing findings on ordination-cardination were not meager enough, a further conceptual problem leads us to doubt the reliability of even this very thin data base. With the exception of the Siegel and Wang-Resnick-Boozer investigations, all of the relevant studies have employed methods of measuring children's grasp of ordination and cardination which are the same as the ones that Piaget originally reported in The Child's Conception of Number. This is extremely unfortunate. In Piaget's original studies, he failed to give even superficial consideration to the crucial question of whether or not his indexes of "ordination" and "cardination" are adequete representations of the logical characteristics of ordinal and cardinal number which we examined in Part I of this book. Piaget simply reports his various tasks to the reader without any clear conceptual analysis or logical rationale (see Piaget 1952: Part II). In view of the fact that Piaget wished to draw conclusions about the development of notions whose origin is logical, this may seem a curious procedure to many readers. However, as those readers who are students of Piaget's work are aware, this is a persistent problem which plagues most of his research. The situation would not be too problematical for us if it happened that, quite by accident, Piaget's procedures for measuring ordination and cardination happened to be the clearest and most obvious techniques for assessing these two concepts. It was noted earlier, however, that this is simply not the case. Generally speaking, the relationship between Piaget's ordination and cardination methods and the logical definitions of ordinal and cardinal number is one of very rough analogy rather than strict deduction.

OPERATIONAL DEFINITIONS

If developmental research involving ordination and cardination is to have any meaning, then we must be reasonably certain that the concepts we are calling "ordination" and "cardination" are precise embodiments of the logical definitions of ordinal and cardinal number, respectively. Therefore, before definitive evidence on the order of emergence of ordination, cardination, and arithmetic can be obtained, we must give some long overdue attention to the measurement definitions of ordination and cardination. We consider this problem now.

First, let us recall certain points discussed in the preceding chapter. By "ordination," we understand, at a minimum, the conceptual counterpart of ordinal number, by "cardination," the conceptual counterpart of cardinal number. If we are to gather data which provide meaningful tests of the three theories discussed in Chapter 6, then we must at least try to insure that our measurements of "ordination" and "cardination" incorporate the elements of the logical definitions of ordinal number and cardinal number, respectively, and as little else as possible.

We begin with the problem of measuring ordination. We saw in Chapter 3 that ordinal number is based on the notion of progression. We also saw that the key element in the logical definition of progression—the thing that all progressions have in common—is the idea of a relation that is both transitive and asymmetrical. Progressions are, in effect, always generated by transitive-asymmetrical relations. The transitivity property of the relations which generate progressions is especially important. In logic, the fact that certain asymmetrical relations are also transitive is called the "minimum ordinal proposition"; it is the one thing which is true of any progression. From a logical point of view, therefore, the minimum conditions for ordinal number are these: three (or more) distinct terms, x, y, and z; some asymmetrical relation R; and the fact that the truth of R_{xz} follows from the truth of R_{xy} and R_{yz}. If these three conditions are met, we have a progression and, hence, an example of ordinal number. These facts indicate that ordination, if it is to be a fairly precise psychological counterpart of ordinal numbers, should be measured roughly as follows: First, obtain a simple transitive-asymmetrical relation which children are familiar with ("longer than," "heavier than," "brighter than"). Children, even preschoolers, know that these relations are unidirectional (asymmetrical). But this fact does not necessarily mean that they also understand ordination. Suppose we have three concrete objects, A, B, and C, between which some common asymmetrical relation obtains (for example, A is longer than B and B is longer than C). Now, we are interested in ordination as a concept, not in percepts of ordinality. Hence, let us suppose the differences between the three objects are not immediately apparent to perception. However, let us also suppose that they can be ascertained by carefully comparing the objects. Suppose we establish for the subject that the asymmetrical relation in question obtains between A and B and between B and C. Can the child infer from these two premises that the relation also obtains between A and C, that is, does the child understand the minimum ordinal proposition? If so, then, at least as far as logic is concerned, he or she grasps ordinal number. Children who possess a stable internal concept of progression—presumably founded on prior perceptions of ordinality—should have very little trouble with problems of the sort just described. On the other hand, children who possess no internal concept of progression or who possess an unstable concept should not be able to solve such problems reliably.

We turn now to cardination. For reasons which we are about to consider, constructing a logically acceptable cardination task is a somewhat more difficult matter than constructing a logically acceptable ordination task. First, however, let us recall that the logical definition of cardinal number is based on the notion of similarity. Similarity, in turn, is defined in terms of the correspondence relation. Two collections contain equally many terms if the correspondence between their respective terms is one-to-one, and two collections contain unequally many terms if the correspondence between their respective terms is one-to-many or many-to-one. If our psychological measure is to be isomorphic with cardinal number, then we must assess subjects' ability to judge the relative manyness of two collections solely on the basis of the type of correspondence that obtains between their respective elements. It follows that we must guard against the possibility that judgments of relative manyness can be based on relations between collections other than correspondence. Since, as was the case for ordination, we are interested in the concept of cardination rather than perceived numerosity, we must examine manyness judgments about collections for which term-by-term correspondence is not suggested by immediate perception. We must avoid, for example, collections of concrete objects between which some obvious functional correspondence obtains of which the subject is aware (socks and shoes, spoons and bowls, bats and balls). The correspondence between the collections to be compared must always be established by the subject himself.

The need to insure that correspondence is the only basis on which relative-manyness judgments can be based poses three problems. The first, the relative density of collections, was discussed in the preceding chapter. When the terms of two concrete collections are arrayed in the same space, density is perfectly correlated with relative manyness. Judgments of relative manyness, therefore, can be based entirely on the density cue, and correspondence is not at all essential. This fact would lead us to consider the possibility of arranging the terms in the collections to be compared in parallel rows. However, as soon as the terms have been arranged in this manner, a second problem arises. A rather substantial body of research has shown that when collections of terms are arranged in parallel rows, children tend to pay very close attention to the relative length of the rows. Hence, once the terms are arranged in parallel rows, we must guard against the possibility that manyness judgments can be based on length cues by insuring that relative manyness is not perfectly correlated with relative length. Assuming we have somehow controlled density and length cues, we still have to deal with the problem of counting. When asked to judge whether two collections have equally many or unequally many terms, it is always possible that children simply count the two collections. Obviously, we should try to preclude this possibility.

With the preceding methodological caveats in mind, we now briefly examine some tasks which take them into account. These tasks will be reported in greater detail in Chapter 8. For now, we only consider how they negotiate the density, length, and counting problems. To begin with, we have several pairs of collections which consist of simple geometric figures: circles and triangles. The cardinal numbers of the various classes are relatively small—no more than 10 terms. To eliminate the counting problem, we do three things. First, we include a large enough number of terms in the collections (between 6 and 10) so that it takes several seconds to count them all. Second, we look for signs of counting behavior when we test our subjects. Third, we specifically ask our subjects not to count. To deal with the density problem, we arrange our collections in parallel rows. To deal with the length problem, we make certain that length is not perfectly correlated with manyness. With regard to the latter two points, consider the six pairs of classes shown in Figure 7. 3. Note that neither relative length nor relative density is perfectly predictive of relative manyness.

A final bothersome point about the measurement of cardination is the trade-off between length and density cues. When two collections to be compared contain unequally many terms, as in pairs A, B, D, and F in Figure 7. 3, it is impossible to eliminate both length and density as bases for correct judgments of relative manyness. If two classes containing unequally many terms are made the same length (thereby

FIGURE 7.3

Set of Stimuli Which Controls for the Correlations between Relative Density and Relative Manyness and between Relative Length and Relative Manyness.

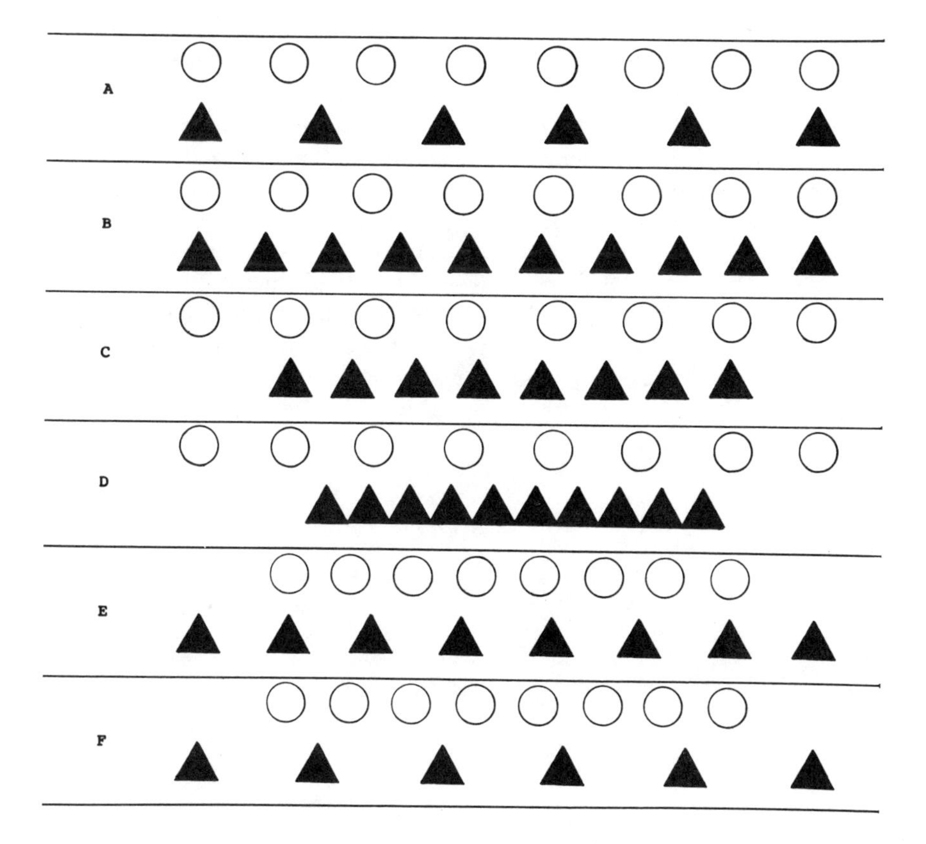

Source: Constructed by the author.

eliminating the length cue), the more populous class automatically becomes the denser of the two (thereby failing to eliminate the density cue). If two classes containing unequally-many terms are made equally dense, on the other hand, the larger class automatically becomes the longer of the two. In short, whenever two collections contain unequally many terms, we cannot be absolutely certain that correct judgments of relative manyness necessitate the use of correspondence. The subject always has at least one other cue in addition to correspondence which is perfectly correlated with relative manyness. Fortunately, we can be certain that neither length nor density cues are employed when two classes contain equally many terms, as in pairs C and E in Figure 7. 3. Whenever two classes contain equally many terms, one of the classes can be made simultaneously longer and less dense than the other, thereby eliminating both cues. This would seem to leave only correspondence as a basis for correct judgments. From this fact, a very important conclusion follows. When it comes to deciding whether or not children possess cardination (defined as understanding the logical connection between correspondence and manyness), judgments about classes containing equally many terms are far more crucial than judgments about classes containing unequally many terms. When a child correctly judges that two classes contain unequally many terms, we cannot be certain that correspondence was the basis for the judgment. In the case of correct judgments of equal manyness, however, correspondence is the only basis for such judgments of which we are aware.

8

THE DEVELOPMENT OF ORDINATION, CARDINATION, AND NATURAL NUMBER

This chapter reports on three large-scale normative studies which were designed to provide empirical evidence bearing on the order in which children acquire ordination, cardination, and natural number. The first of these studies focused exclusively on the problem of whether or not ordination and cardination emerge in a fixed order. The second study was designed to determine whether or not there is a developmental lag between arithmetic proficiency, on the one hand, and ordination-cardination, on the other. These two studies were conducted in the author's laboratories in Edmonton during 1972. The subjects participating in the studies were Canadian kindergarten and first-grade children. The final study reported in this chapter was an independent project conducted by another investigator during late 1973 and early 1974. Its aim was to determine whether the principal findings of the first two studies reported in this chapter could be replicated in another laboratory using the same methods but different subjects. The subjects were kindergarten and first-grade children who resided near Madison, Wisconsin.

STUDY 1

This initial investigation was designed to focus narrowly on the order in which ordination and cardination emerge in the reasoning of a large sample of kindergarten and first grade children. As will be seen presently, the tasks were based on the operational definitions of "ordination" and "cardination" given at the end of the preceding chapter. Kindergarden and first-grade children were studied for essentially psychometric reasons. When one investigates the order in which two or more concepts emerge in children's thinking, it is important to choose an age level at which children presumably will show some facility with both concepts. If the subjects are too young, they will not possess any of the concepts being measured and no sequence will be observed in the data. If the subjects are too old, they will possess all the concepts and again no sequence will be observed. Therefore, it is desirable to study intermediate age levels at which some variability in both concepts can be expected. Previous research on ordination and cardination, particularly Piaget's landmark studies, indicated that the kindergarten-first grade level would be an appropriate place to begin.

Method and Procedure

Subjects

A total of 180 children served as subjects in this study. Kindergarteners and first graders each made up 50 percent of the sample, and both halves of the sample consisted of equal numbers of boys and girls. The average age of the 90 kindergarteners was five years, nine months, and the average age of the 90 first graders was six years, eight months. The children who participated in the study were selected from the class lists of five different elementary schools. All of the schools were located in middle-class areas of the city of Edmonton.

Language Pretests

Developmental researchers have known for some time that logical and mathematical concepts frequently develop independently of the capacity to use these concepts linquistically (Brainerd 1973d, 1973e). Consequently, when tests for such concepts involve understanding them linguistically, as the present ordination and cardination tests do, it is possible for children who actually possess the concepts in question to fail the tests. This is usually called a "false negative" measurement error. To eliminate such errors, it is traditional to pretest children for their understanding of the terminology to be employed in the concept tests, and such a procedure was employed in this study. In particular, the subjects' grasp of certain key "relational terms" was assessed before the ordination and cardination tests were administered. There were two separate pretests for relational terms employed in the ordination tests: a length pretest and a weight pretest. In the first ordination pretest, the subject was shown a picture of two parallel lines measuring three inches and six inches in length, respectively, and was asked three questions about the picture: (1) Are these lines the same length? (2) Is one of these lines longer (and if so, which one)? (3) Is one of these lines shorter (and if so, which one)? In the second ordination pretest, two clay balls weighing 16 ounces and 24 ounces, respectively, were placed in the subject's hands and he was asked: (1) Do the balls weigh the same? (2) Does one of the balls weigh more? (and if so, which one)? If the subject answered all of these questions correctly, then it was concluded that he or she had an adequate grasp of the terminology to be employed on the ordination tests. Any subject failing to answer any of these questions correctly was not allowed to participate in the remainder of the study.

The cardination language pretests consisted of two story problems. The first story problem ran as follows: "Suppose that I had eight cookies and you had six cookies. Would one of us have more cookies (and if so, whom)? Would one of us have fewer cookies (and if so, whom)? Would we both have the same number of cookies?" The second story problem ran as follows? "Suppose that I had four cookies and you had four cookies. Would one of us have more cookies (and if so, whom)? Would one of us have fewer cookies (and if so, whom)? Would we both have the same number of cookies?" If the subject answered all of the questions on both story problems correctly, then it was concluded that he or she had an adequate grasp of the terminology to be employed on the subsequent cardination tests.

Procedure for the Ordination Tests

All 180 subjects were administered two different tests of ordination: a length test and a weight test. During both tests, the subject and the experimenter sat across from each other at a small table. There were five steps in the length-ordination test. First, the experimenter placed a large board in the center of the table. Two sticks were glued in parallel on the board with a distance of two feet between them. The sticks appeared to be the same length, but they actually differed by three-eighths of an inch. (One of the sticks was 11-inches long and the other 11 and three-eighths inches.) Both sticks glued to the board were colored red. Second, the experimenter produced a yellow measuring stick which was not glued to the board. This stick appeared to be the same length as the two sticks glued to the board; however, it was three-sixteenths of an inch longer than one and three-sixteenths shorter than the other. Third, the experimenter placed the measuring stick beside the 11-inch stick, and asked the subject which of the two sticks was longer. Fourth, the experimenter placed the measuring stick beside the 11 and three-eighths inch stick, and asked the subject which of the two sticks was longer. Fifth, to determine whether or not the subject understood that the asymmetrical relation "longer than" is also transitive, the experimenter posed the following questions: (1) Are the two red sticks the same length? (2) Is one of the red sticks longer then the other (and if so, which one)? (3) Is one of the red sticks shorter than the other (and if so, which one)? After the subject had answered each of these questions, the ordination of length procedure was repeated in the reverse direction. That is, this time the measuring stick was placed beside the 11 and three-eighths inch stick first and beside the 11-inch stick second. Once these two comparisons had been made, the preceding three questions were repeated.

The second ordination test, ordination of weight, also involved five steps. First, the experimenter placed a large board in the center of the table on which three green plastic plates were glued in a row. There was an eight-ounce red clay ball in one of the outside dishes and a 24-ounce red clay ball in the other outside dish. The two balls were exactly the same size. The middle dish glued to the board was empty. Second, the experimenter introduced a yellow measurement ball weighing 16 ounces and placed it in the empty center dish. Third, the experimenter placed the eight-ounce ball in the subject's left hand, placed the measurement ball in the subject's right hand, and asked the subject to judge which of the two was the heavier. Fourth, the experimenter shifted the measurement ball to the subject's left hand, placed the 24-ounce ball in the subject's right hand, and asked the subject to judge which of the two was the heavier. Fifth, to determine whether or not the subject understood that the asymmetrical relation "heavier than" is also transitive, the experimenter posed three questions: (1) Do the two red balls weigh the same? (2) Does one of the red balls weigh more (and if so, which one)? (3) Does one of the red balls weigh less (and if so, which one)? As soon as all three questions had been answered, the ordination of weight procedure was repeated in the reverse direction. That is, the measuring ball was compared with the 24-ounce ball first and with the eight-ounce ball second. Once the two comparisons had been made in the reverse direction, the preceding three questions were posed again.

Procedure for the Cardination Tests

As was the case for the ordination tests, the experimenter and the child sat across from each other at a small table during the cardination tests, and there were five steps involved. First, the experimenter showed the subject a picture of a row of eight evenly spaced red dots. Second, the following instructions were given to the subject: "We are going to play a game with some picture cards like this one. All the picture cards have a row of red dots like this one. All the picture cards also have a row of blue dots. This is how we play the game. First, I show you a picture. You take a good look at the picture and try to figure out whether there are the same number of red and blue dots or whether one of the two rows has more dots. We have a special rule in this game that makes it more fun: You cannot count any of the dots—you have to figure out the answer some other way." The "special rule," of course, was to eliminate the counting problem mentioned at the end of Chapter 7. Third, the experimenter introduced one of the six test pictures shown in Figure 8.1. In these test pictures, the upper row of dots was always red and the lower row was always blue. Concerning the problems of length and density mentioned at the end of Chapter 7, note that length is never predictive of relative manyness and that density is not predictive of relative manyness in pictures C and E. Two questions were posed in conjunction with each of the six pictures: (1) Are there just as many red dots as there are blue dots in this picture? (2) Does one of the rows have more dots than the other row (and if so, which one)? Fourth, each of the remaining five test pictures was presented, and the experimenter posed the two questions just described in conjunction with each picture. Fifth, to further reduce the possibility that correct judgments of relative manyness were based on counting, two additional checks were incorporated in the cardination tests. First, after the subject had seen all the pictures and answered all the questions, the experimenter asked, "Did you find that you could answer all the questions without counting the dots or did you have to count sometimes?" Second, the experimenter who administered the cardination tests was trained to watch for rhythmic hand movements, arm movements, lip movements, and other overt motor signals that covert counting was going on. The latter check is an exceptionally effective one because, below the age of eight or nine years, most children find it virtually impossible to count without some supporting motor response. If, in answer to the aforementioned question, the subject said that he or she had never counted, the experimenter then asked him or her to explain what method had been employed to judge relative manyness. Any subject who either admitted having counted or had been judged as having counted by the experimenter, was dropped from the study. A total of 11 children (8 first graders and 3 kindergarteners) said that they had counted. The experimenter had observed lip counting and pointing in all 11 subjects.

Principal Findings of the Study

As we have just seen, each subject made a total of 12 judgments on both the ordination and cardination tests. On both tests, the subject was given one point for each correct judgment.

Development of Ordination

When the children's performance on the ordination tests was examined, it was noted that they tended to fall into three general categories. Moreover, the three categories tended to be correlated with age, that is, subjects falling in the first category tended to be the youngest ones in the sample, subjects falling in the second category

FIGURE 8.1

The Six Stimuli Employed in the Cardination Tests

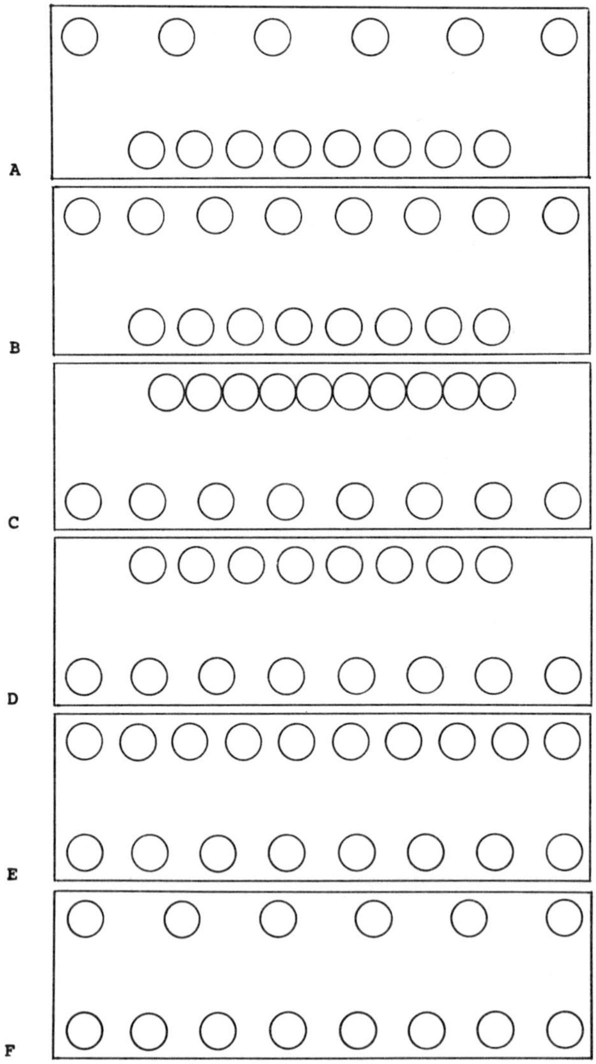

Source: Constructed by the author.

tended to be somewhat older, and subjects falling in the third category tended to be the oldest subjects in the sample. The important features of the three categories will be briefly summarized.

Level I: No ordering. These subjects evidenced no understanding that the everday asymmetrical relations "longer than" and "heavier than" are invariably transitive. When the experimenter asked them about the relative length of the two red sticks or the relative weight of the two red balls, these subjects tended to answer that the two were equivalent. That is, they appeared to accept their perceptual impression of the two objects rather than to construct an ordered progression on the basis of the information which they had been given. The average subject in this category did not answer any of the 12 ordination questions correctly. The average age of the subjects in this category was four years, eleven months.

Level II: Partial ordering. These subjects understood that "longer than" and "heavier than" are transitive in some cases, but they did not understand it in other cases. Explicitly, they tended to answer the questions correctly on the first half of each of the two ordination tests, when the ordering was left-to-right, but they tended to answer the questions incorrectly when the ordering was reversed on the second half of each test. Also, on the second and third questions mentioned above, it was common for these subjects to give a correct "no" judgment and then not be able to point out the longer and shorter of the two red sticks or the heavier and lighter of the two red balls. These facts suggest that the subjects' ordering of the three terms was not a completely internal one. Instead, the relative length or relative weight of the three terms was confused with their spatial positions on the board. The average subject in this category answered eight of the ordination questions correctly. The average age of the subjects in this category was five years, three months.

Level III: Complete internal ordering. These subjects understood that "heavier than" and "longer than" are always transitive. Unlike Level II subjects, Level III subjects encountered no difficulty with the second halves of the two ordination tests. They also were able to point out the longer and shorter of the two red sticks and the heavier and lighter of the two red balls on the second and third questions. Thus, their ordering of the terms in each progression appeared to be a completely internal one. The average subject in this category answered all 12 of the ordination questions correctly. The average age of these subjects was five years, eight months.

The ordination-test performance of all 180 subjects was reviewed in light of the three levels of ordination just described. Each subject was classified as belonging to one and only one of the levels: a subject answering five or fewer questions correctly was assigned to level I; a subject who answered less than eleven but more than five questions correctly was assigned to level II; and those who answered either eleven or twelve questions correctly were assigned to level III. The various numbers of kindergarteners and first graders assigned to each category are shown in Table 8.1 which clearly illustrates the age-relatedness of the classification scheme. More kindergarteners than

TABLE 8.1

Numbers of Kindergarten and First-Grade Children Assigned to Each of the Three Ordination Levels

Age	Level I	Level II	Level III
Kindergarten	24	30	36
First grade	6	27	57

Source: Compiled by the author.

first graders were assigned to level I, and more first graders than kindergarteners were assigned to level III. (Chi-square tests of statistical significance indicated that there was less than one chance in 1,000 that the former difference is due to chance alone and that there is less than one chance in 500 that the latter difference is due to chance alone.) Age-relatedness also is illustrated by the fact that when the first graders are divided into "younger" and "older" halves, 19 of the 27 first graders assigned to level II fell in the younger group. (A chi-square test indicated that there was less than one chance in 97.5 that this age difference is due to chance alone.)

In summary, an examination of the children's performance on the ordination tests produced a three-level classification scheme. The three levels of the scheme were correlated with age in such a manner as to suggest that they may be conveniently viewed as representing levels, albeit somewhat arbitrary levels, in the emergence of ordination in children's reasoning.

Development of Cardination

When the children's cardination test scores were examined, it was noted that (1) cardination performance, like ordination performance, tended to fall into three categories, and (2) the three cardination categories were definitely correlated with age.

Level I: No correspondence. Subjects falling in this initial category showed no evidence of cardinal number. That is, there was no indication that they understood that the type of correspondence that obtained between the two rows of dots always determined their relative manyness. These subjects' judgments appeared to be based exclusively on the relative lengths of the two rows. Whenever the two rows were of unequal length (pictures C, D, E, and F in Figure 8.1), the longer row was judged to contain more dots than the shorter row. Whenever the rows were the same length (pictures A and B), they were judged to contain the same number of dots. The average subject in this category answered two of the cardination questions correctly. The average age of the subjects in this category was five years, ten months.

Level II: No one-to-one correspondence. Subjects falling in this category clearly did not base their manyness judgments on the length cue. They appeared to be able to judge unequal manyness correctly, but not equal manyness. Whenever the two rows contained unequally many dots (pictures A, B, D, and F in Figure 8.1) and regardless of the lengths of the two rows, these subjects tended to judge relative manyness correctly. However, whenever the two rows contained equally many dots (pictures C and E in Figure 8.1), these subjects incorrectly judged the shorter row as the more numerous. Two possible conclusions are possible about these subjects' manyness judgments. First, it could obviously be argued that they understand the logical connection between relative manyness and correspondences of one to many and many to one, but they have not yet worked out the connection between one-to-one correspondence and relative manyness. On the other hand, it could also be argued that these subjects are simply basing their judgments on perceived density and do not understand anything about the connection between correspondence and relative manyness. The fact that they tend to judge the shorter (and denser) row as more numerous whenever the two rows contain equally many dots strongly suggests that the second interpretation of how level II subjects generate their manyness judgments probably is the correct one. Also, when they were asked to explain how they arrived at their judgments, these subjects said things such as, "It depends on how close things are together." The average subject in this category answered eight of the cardination questions correctly. The average age of the subjects in this category was six years, eight months.

Level III: Complete internal correspondence. Subjects falling in this category appear to understand fully the logical connection between correspondence and relative manyness. Regardless of whether the two rows contain equally many or unequally many dots, they tend to judge their relative manyness correctly. These subjects appear to be oblivious to the length and density cues on which level I and level II subjects seem to be completely dependent. The average subject in this category answered all twelve cardination questions correctly. The average age of the subjects in this category was seven years, one month.

The cardination-test performance of the 169 children who did not count, or at least for which there was no evidence of counting, was reviewed in light of the three categories just described. Each subject was classified as belonging to one and only one of the categories via the following criteria: if five or fewer questions were answered correctly, the subject was assigned to level I; if less than eleven but more than five questions were answered correctly, the subject was assigned to level II; and, if eleven or twelve questions were answered correctly, the subject was assigned to level III. The subjects appear by level in Table 8.2. As was the case for the ordination classification scheme, the age-relatedness of the cardination levels is clearly evident. There appear to be more kindergarteners than first graders at level I, more first graders than kindergarteners at level II, and more first graders than kindergarteners at level III. (Chi-square tests indicated that there is less than one chance in 1,000 that the first difference may be attributed to chance alone, less than one chance in 95 that the second dif-

TABLE 8.2

Numbers of Kindergarten and First-Grade Children Assigned to Each of the Three Cardination Levels

School level	Level I	Level II	Level III
Kindergarten	59	25	2
First grade	33	37	13

Source: Compiled by the author.

ference may be attributed to chance alone, and less than one chance in 500 that the third difference may be attributed to chance alone.) Thus, we have a cardination classification scheme which, like the aforementioned ordination scheme, is correlated with age in a manner which suggests that the three categories may be viewed as representing levels of increasing understanding of cardination.

The Developmental Relationship Between Ordination and Cardination

Now, let us take up the problem which this study was expressly designed to answer—whether or not there is developmental lag between ordination and cardination. There is a fairly simple way to go about answering this question with these data. It involves comparing the ordination level and cardination level of each subject in the sample to see if, on the average, there is a discrepancy between the two. Each subject had already been assigned to one and only one of the three ordination levels and to one and only one of the three cardination levels. To effect a comparison of ordination and cardination, it is only necessary to cross-classify the subjects according to both their levels of cardination and their levels of ordination. The ordination and cardination performance of the 169 subjects who did not count were cross-classified in this manner, and the results appear in Table 8.3 by ordination level and cardination level. Even a brief glance at this table indicates that there is very pronounced developmental lag between ordination and cardination. Of the 93 subjects in the present sample who always understood that asymmetrical relations are also transitive (level III of ordination), almost one-half (40) did not evidence the slightest understanding of the fact that the relative manyness of two collections depends on the type of correspondence which obtains between

TABLE 8.3

The Developmental Relationship Between Ordination and Cardination

Level of Ordination	Level of cardination		
	I	II	III
I	20	4	0
II	32	17	3
III	40	41	12

Source: Compiled by the author.

them (level I of cardination). Conversely, most (72) of the 92 children assigned to level I of cardination were capable of either partial internal ordering (level II of ordination) or complete internal ordering (level III of ordination). Statistical tests of the significance of these findings are carried out via the following elementary procedure. First, the total number of subjects who were functioning at a higher level of ordination than cardination is determined (Table 8.3 indicates 113 in all). Second, the total number of subjects who were functioning at a higher level of cardination than ordination is determined (seven). Third, these two values are added to yield the total number of subjects whose ordination and cardination levels were discrepant (120). Fourth, a ratio is constructed with the first value (113) in the numerator and the third value (120) in the denominator. Fifth, the ratio is tested for statistical significance by employing the binomial formula $p(x) = \binom{N}{x} p^x Q^{N-x}$, where p(x) is the probability of obtaining the ratio by chance alone, x is the number of subjects functioning at a higher level of ordination than cardination, and N is the total number of subjects. Both p and Q are set at 0.50. The computation indicates that there is less than one chance in one billion that the ratio 113/120 could have occurred by chance alone.

Generally speaking, therefore, the first study provided strong support for the ordinal theory's prediction that children grasp ordination before they understand the connection between correspondence and manyness. The contrasting developmental predictions of the cardinal and cardinal-ordinal theories appear correspondingly untenable. The children who participated in this investigation evidenced sophisticated and completely internalized ordering long before they displayed comparable facility with cardination. In fact, as is apparent from Table 8.3, it was common for children functioning at the highest level of ordination to evidence absolutely no understanding of the connection between correspondence and relative manyness.

STUDY 2

The second of our three normative studies was undertaken for three general reasons. First, the data of the initial study clearly indicated that ordination is developmentally prior to cardination. However, before we conclude that ordination emerges first, prudence dictates that we at least try to replicate the findings of study 1 in a new group of subjects. Second, this study was designed to determine whether or not there is a developmental lag between ordination and natural-number competence. Third, it was also designed to determine whether or not there is a developmental lag between cardination and natural-number competence.

To address these questions, a new sample of kindergarten and first-grade children was selected. This time arithmetic proficiency was measured as well as ordination and cardination. Arithmetic proficiency was viewed as the best general estimator of children's natural-number competence. The ordination and cardination tests were the same as the ones just discussed. Two methods of measuring arithmetic competence were considered. To begin with, a combination of the children's report-card grades and their teachers' opinions was considered. This method was rejected for the following reasons. It is unfortunate but true that different schools and different teachers often have very different views about what is and is not "arithmetic competence." In many elementary schools, for example, report-card grades in arithmetic are based in large part on children's grasp of elementary physical concepts and elementary geometric ideas. Now, although physical and geometric notions are of considerable practical importance and certainly merit instruction, they obviously do not have anything to do with arithmetic considered as a formal mathematical system. In short, given the many other things which are incorporated in elementary school "arithmetic" other than arithmetic itself, making inferences about the developmental relationship between ordination-cardination and arithmetic competence on the basis of grades and teacher opinion would seem to be hazardous at best. An even more important problem with report-card grades and teacher opinion is the well-known fact that these criteria always incorporate—to a greater or lesser extent depending on the school and the teacher—irrelevant social and personal characteristics of pupils.

Hence, if we wish to avoid the criticism that what we are calling arithmetic proficiency is not what we say it is, it behooves us to make certain that the variables we measure are such as would be called "arithmetic" by everyone. To do this, a sample of subjects was selected from five elementary schools whose arithmetic programs were, in fact, concerned with arithmetic in the classical sense of the term. During kindergarten, the schools focused on teaching children to add the first few positive integers ($1 + 1 = 2$, $1 + 2 = 3$, . . . , $4 + 4 = 8$). During first grade, the schools continued to teach addition of the first few positive integers and, in addition, to teach subtraction of these same integers ($2 - 1 = 1$, $3 - 1 = 2$, . . . , $8 - 4 = 4$). Therefore, it was a simple matter to determine the present subjects' arithmetic proficiency by administering individualized tests of addition and subtraction. A combined verbal and written method was employed. For example, to test a subject's understanding of $1 + 3 = 4$, the subject would be presented with a card on which the symbols "$1 + 3 = ?$" were printed. Next, the experimenter would ask, "How many apples are one apple plus three apples?" The child who either wrote the symbol "4" or said "four," or did both, was judged to have passed the item.

Method and Procedure

Subjects

As was the case for study 1, 180 kindergarten and first-grade children participated in this study. There were equal numbers of subjects from each age level, and there were equal numbers of boys and girls. The subjects were drawn from the class lists of five middle-class elementary schools in Edmonton. These schools had been previously selected on the basis of their arithmetic programs. In all the schools, arithmetic instruction at the kindergarten and first grade levels focused on teaching children how to add and subtract the first few positive integers. The average age of the 90 kindergarten participants was 5 years, 10 months. The average age of the 90 first grade participants was 6 years, 10 months. The same language, ordination, and cardination tests administered to the children who participated in study 1 also were administered to the children who participated in this study.

Natural Number

From a measurement standpoint, the distinguishing feature of this study is that the subjects were administered comprehensive tests of their ability to add and to subtract the first few natural numbers. During these tests, the experimenter and the child sat across from each other at a small table. The natural-number test consisted of 32 problems in all. A total of 16 problems each were concerned with addition and subtraction. Because the kindergarten children who participated in this study had not yet received any systematic instruction in subtraction they were administered only the 16 addition items of the natural-number test. However, the first-grade subjects, who had already received instruction in both addition and subtraction, were administered all 32 items.

At the beginning of the test, the experimenter gave some brief instructions to the subject. In the case of the kindergarteners, the subjects were informed that they were going to play a game in which the subjects' task was to add up some numbers and in the case of the first graders, subjects were informed that they were going to play a game in which their task was both to add and to subtract some numbers. The subjects then were given pencils. At the beginning of each test item, the experimenter placed a sheet of paper on which the problem was printed in standard numerical symbols in front of each subject. The experimenter then repeated a story problem concerning the addition or subtraction of apples which was the same as the problem printed on the sheet. After the experimenter had posed the problem verbally, the child was allowed to respond by saying the answers aloud, by writing it down on the sheet of paper, or by doing both. The 16 specific addition items and the 16 specific subtraction items which made up the test are shown in Table 8.4. Two brief illustrations will clarify the manner in which the items were administered. Suppose the item was the third addition problem shown in Table 8.4. To begin with, a sheet would be placed in front of the subject on which the symbols $1 + 3 = ?$ would be printed. Once the subject had looked at the symbols on the sheet, the experimenter would ask, "How many apples are one apple plus three apples?" If the subject said "four," wrote "4" on the sheet, or did both, then he or she was judged to have passed the item. Second, suppose the item was the fourth subtraction problem shown in Table 8.4. A sheet would first be placed in front of the subject on which $5 - 1 = ?$ would be printed. After the subject had looked at the symbols the experimenter would ask, "How many apples are five apples minus one apple?" If the subject said "four," wrote "4" anywhere on the sheet, or did both, the item was passed.

Principal Findings of the Study

Ordination and Cardination

In study 1, the subjects appeared to grasp ordination long before they understood the relationship between correspondence and relative manyness. In view of the importance of this finding from the standpoint of our three theories of number development, the first step in analyzing the data of study 2 was to determine whether ordination also preceded cardination in the present group of subjects. To answer this question, the subjects' performance on the ordination and cardination tests were scored in the same manner as described earlier. After this was done, it was noted that the three general findings of study 1 also held for the present subjects: (1) the same three levels of ordination development were observed, (2) the same three levels of cardination development were observed, and (3) most importantly, when the subjects' ordination and cardination performances were compared, ordination was again observed to be far in advance of cardination. For 98 of the 180 subjects, the level of ordination to which they had been assigned was either one or two levels higher than the level of cardination to which they had been assigned. In contrast, only 10 subjects were assigned to a higher level of cardination than ordination.

Developmental Relationships between Ordination, Cardination, and Natural Number

We turn now to the two new questions that study 2 was designed to answer: Is there a developmental lag between ordination and the

TABLE 8.4

Individual Items on the Test of Arithmetic Proficiency

Test/Item	Problem
Addition/1	1 + 1 = ?
Addition/2	1 + 2 = ?
Addition/3	1 + 3 = ?
Addition/4	1 + 4 = ?
Addition/5	2 + 1 = ?
Addition/6	3 + 1 = ?
Addition/7	4 + 1 = ?
Addition/8	2 + 3 = ?
Addition/9	2 + 4 = ?
Addition/10	3 + 2 = ?
Addition/11	4 + 2 = ?
Addition/12	3 + 4 = ?
Addition/13	4 + 3 = ?
Addition/14	2 + 2 = ?
Addition/15	3 + 3 = ?
Addition/16	4 + 4 = ?
Subtraction/1	2 - 1 = ?
Subtraction/2	3 - 1 = ?
Subtraction/3	4 - 1 = ?
Subtraction/4	5 - 1 = ?
Subtraction/5	5 - 2 = ?
Subtraction/6	6 - 2 = ?
Subtraction/7	7 - 3 = ?
Subtraction/8	8 - 4 = ?
Subtraction/9	2 - 1 = ?
Subtraction/10	3 - 2 = ?
Subtraction/11	4 - 3 = ?
Subtraction/12	5 - 3 = ?
Subtraction/13	5 - 4 = ?
Subtraction/14	6 - 4 = ?
Subtraction/15	7 - 4 = ?
Subtraction/16	8 - 4 = ?

Source: Compiled by the author.

natural-number concept? Is there a developmental lag between cardination and the natural-number concept? Before it was possible to examine these questions, the subjects' performance on the addition and subtraction problems had to be scored. To facilitate the comparison of the development of arithmetic competence with the development of ordination and cardination, the addition scores of all 180 subjects and the subtraction scores of the 90 first-grade subjects were grouped into three somewhat arbitrary categories: below average, average, and above average. To determine what, generally speaking, is below average, average, and above average arithmetic proficiency for children of this age level, the teachers and principals of the schools participating in the study were interviewed and their thoughts on the matter were requested. On the basis of these interviews, the following criteria were formulated and subsequently were employed to assign subjects to the aforementioned categories: children passing five or fewer of the addition items were assigned to the below average addition category, those passing more than five but less than 12 of the addition items were assigned to the average addition category, and those who passed between 12 and 16 of the addition items were assigned to the above average addition category. For the sake of consistency, the same cutoff points were used to assign the 90 first graders to below average, average, and above average subtraction categories.

To determine whether or not there are developmental lags between arithmetic proficiency and ordination-cardination, the subjects were cross-classified as they were in study 1. This time, however, they were cross-classified in terms of three variables rather than two: that is, level of ordination, level of cardination, and level of arithmetic performance. The results of the classification appear in Table 8.5 by ordination level, cardination level, and arithmetic level. Inspection of Table 8.5 indicates that there appear to be developmental lags between both of the former variables and arithmetic. Taking ordination first, children's grasp of this concept appears to precede both addition and subtraction. A total of 90 of the 180 subjects were assigned to a higher level of ordination than addition, while only eight were assigned to a higher level of addition than ordination. Note, in particular, that more than one-third of the former group were functioning at the highest level of ordination but were still below average in addition. Turning to subtraction, a total of 31 of the 90 first graders were assigned to a higher level of ordination than subtraction, while only 3 were assigned to a higher level of subtraction than ordination. Moreover, slightly more than 50 percent of the subjects functioning at a higher level of ordination were assigned to the highest level of ordination and the lowest level of subtraction. Thus, children appear to grasp ordination before they begin making very much progress with arithmetic. There also is a developmental relationship between cardi-

TABLE 8.5

The Developmental Relationships Between Ordination-Cardination and Arithmetic Competence

Arithmetic	Level of ordination			Level of cardination		
Competence	I	II	III	I	II	III
Addition						
Below average	20	17	35	55	10	0
Average	2	16	38	30	14	7
Above average	0	6	46	25	19	20
Subtraction						
Below average	7	14	16	18	3	0
Average	0	6	17	8	16	6
Above average	0	3	27	10	21	8

Note: Figures for subtraction competence include only the 90 first graders in study II.

Source: Compiled by the author.

nation and arithmetic proficiency, but it is precisely the reverse of the one we have just noted for ordination. First, consider cardination and addition. A total of 64 of the 180 subjects were assigned to a higher addition level than cardination level, and only 12 were assigned to a higher level of cardination than addition. Similarly, 29 of the first graders were assigned to a higher subtraction level than cardination level, and only seven were assigned to a higher level of cardination than subtraction. Thus, the children who participated in this experiment appeared to make considerable progress with both addition and subtraction before they understood cardination.

STUDY 3

Taken together, study 1 and study 2 seem to provide clear and consistent support for the developmental predictions of the ordinal theory. Both studies also would appear to invalidate the contrasting predictions of the cardinal theory and Piaget's cardinal-ordinal theory. The cardinal theory posits—concerning the origin of the natural-number concept—that understanding the connection between correspondence and relative manyness is a necessary precondition for arithmetic. However, many of the subjects who participated in study 2 were able to perform at well above average on both addition and subtraction problems even though they did not show the slightest understanding of the connection between the relative manyness of two collections and the type of correspondence between them. In contrast, those subjects in study 2 who showed clear evidence of cardination invariably performed at above average on the addition and subtraction problems. These facts suggest that whatever else may be true of the relationship between arithmetic proficiency and cardination, it does not seem that one must grasp cardination before one can learn how to add and subtract. These facts are also inconsistent with Piaget's theory of number development (Piaget 1952, Beth and Piaget 1966). Piaget, it will be recalled, believes that ordination and cardination are both necessary preconditions for arithmetic. The present data confirm only the first of these predictions.

In science, of course, the ultimate test of the validity of any set of findings is their replicability. If other investigators—working in different laboratories using more or less the same procedures—cannot reproduce the findings, then this fact usually carries the automatic implication that the original data are, somehow, unique to the laboratory or the investigator who obtained them. Unique and nongeneral results of this sort are not of the slightest interest to science. Given the centrality of the replicability criterion, it is important to ask whether anyone has been able to replicate the findings reported above. Although a substantial number of replication studies has yet to accumulate, a few studies have been conducted and reported. Importantly, there appears to be no investigator who has attempted to replicate the preceding findings, via remotely comparable procedures, who has not been able to do so.

There exists one replication study in particular, the findings of which merit special attention, both because of its comprehensiveness and the comparability of its procedures to those described above. Among other things, the study was designed to determine whether or not the developmental lags between ordination, cardination, and arithmetic reported above could be replicated. As noted earlier, studies 1 and 2 were conducted during 1972. The principal results became known shortly thereafter to other investigators. During 1973, a group headed by F. H. Hooper at the University of Wisconsin's Research and Development Center for Cognitive Learning became interested in the replicability of these results. They decided to undertake a systematic replication attempt as part of a much larger project dealing with the development of children's logical and scientific concepts. The investigation was conducted during late 1973 and early 1974 by A. Gonchar (1975).

TABLE 8.6

The Developmental Relationship Between Ordination and Cardination for the Wisconsin Sample

Ordination Level	Cardination Level		
	I	II	III
Kindergarten			
I	9	0	0
II	18	3	0
III	23	7	0
Third grade			
I	1	0	0
II	0	11	3
III	6	26	13

Source: From Gonchar (1975).

The subject sample for the replication consisted of 60 kindergarten children and 60 third-grade children. All the subjects were enrolled in one of four elementary schools located in Beloit, Wisconsin. There were 30 boys and 30 girls at each age level. The average age of the kindergarteners was five years, ten months and the average age of the third graders was eight years, eleven months. There were two minor differences between the replication investigation and the procedures reported for studies 1 and 2. First, the subjects received three testing sessions rather than one. Second, the subjects were administered several additional tests of relational and classificatory concepts plus a memory test. The important point, however, is that all subjects received the ordination, cardination, and natural-number tests described earlier.

After the testing sessions had been completed, the classification schemes employed in studies 1 and 2 were used to assign each subject to one of the three levels of ordination, one of the three levels of cardination, and one of the three levels of arithmetic proficiency. Next, the subjects were cross-classified as before to determine whether there were any developmental lags. The principal findings for the total sample are these. Concerning the sequence of ordination preceding cardination observed in study 1 and in study 2, 78 of the subjects in the replication sample were functioning at higher levels of ordination than cardination whereas only three subjects were functioning at higher levels of cardination than ordination. A total of 29 of the former subjects were functioning at the highest level of ordination and the lowest level of cardination. Concerning the sequence of ordination preceding natural number observed in study 2, 40 of the replication subjects evidenced higher levels of ordination than arithmetic whereas only 16 subjects evidenced the reverse pattern. Finally, concerning the sequence of natural number preceding cardination observed in study 2, 58 of the replication subjects evidenced higher levels of arithmetic proficiency than cardination whereas only four subjects evidenced the reverse pattern. A more comprehensive report of the findings is given by age level in Tables 8.6 and 8.7.

Obviously, these findings do not differ from those gathered in the author's laboratories in any major respect. On the contrary, the two sets of findings are in almost complete agreement. Table 8.6 indicates that the Wisconsin children, like their Canadian counterparts, understood ordination long before they understood cardination. Table 8.7 indicates, first, that the Wisconsin children also understood ordination before they made much progress with arithmetic and, second, that they had made considerable progress with arithmetic before they

TABLE 8.7

The Developmental Relationship Between Ordination, Cardination, and Arithmetic for the Wisconsin Sample

Arithmetic Level	Ordination level			Cardination level		
	I	II	III	I	II	III
Kindergarten						
I	8	12	16	33	3	0
II	0	5	10	10	5	0
III	0	4	4	7	2	0
Third grade						
I	0	0	1	1	0	0
II	0	0	1	0	4	1
III	1	10	43	7	33	15

Source: From Gonchar (1975).

evidenced much understanding of cardination. The agreement between these findings for the Wisconsin sample and the findings mentioned earlier is, it seems, sufficiently obvious to require no further comment.

The Wisconsin replication produced one other result which merits mentioning. In addition to the analysis procedures discussed in conjunction with study 1 and study 2, Gonchar also examined a very conservative method of identifying developmental lags between ordination, cardination, and natural number based on Piaget's theory. The sizes of the developmental lags were reduced when the conservative method was used. Importantly, however, this does not constitute a failure to replicate, because there are simple mathematical reasons for this finding concerned with the effects of false negative measurement errors. A discussion of these reasons falls far outside the scope of our inquiry. Readers who are especially interested in this problem are directed to appropriate papers in the technical literature dealing with the effects of measurement errors in concept development research (Bingham-Newman and Hooper 1975; Brainerd 1975a, 1975b, 1976; Brainerd and Hooper 1975).

9

TRAINING ORDINATION AND CARDINATION

Although the normative investigations reported in the preceding chapter provide consistent support for the predictions of the ordinal theory of number development, the data examined dealt exclusively with the developmental predictions of the theory. We know from Chapter 6 that the theory makes other testable predictions. In particular, it predicts that when children first acquire basic arithmetic skills, there is a close functional connection between their grasp of ordination and their progress in arithmetic. Other things being equal, children who are capable of completely internalized ordering will make more rapid progress with arithmetic than children who are incapable or only partially capable of internal ordering. The ordinal theory also predicts that there is little, if any, connection between children's understanding of cardination and the emergence of first arithmetic skills. Assuming normal development in the ordination sphere, knowing that the relative manyness of two collections depends on the type of correspondence between them will not necessarily determine competence in the initial facts of arithmetic. The ordinal theory explains these two predictions as follows: To begin with, it is posited that, in their natural environments, when preschool and early elementary school children first learn concrete meanings for the written numerals "1," "2," "3," . . . and the spoken numerals "one," "two," "three," . . . (or for whatever other written and spoken number symbols a given culture uses), they overwhelmingly tend to learn an ordinal meaning.* Thus, "2" and "two" usually are first taken to mean "second" rather than "pair," "3" and "three" are first understood as

---

*Note that it is not suggested that <u>all</u> children learn the ordinal meaning first. It is only contended that the great preponderance of them do.

"third" rather as "trio," and so on for at least the first five or six natural numbers.

In truth, the preceding explanation of the functional relationships between ordination-cardination and acquiring arithmetic skills is itself only an empirical prediction. Even if the prediction were confirmed by hard evidence, we still would not know why numerals are first assigned their ordinal meaning rather than their cardinal meaning. After all, as we saw in Chapter 6, the everyday environment seems to provide numerous examples of both the property of order and the property of manyness. What causes children to choose the former over the latter? The ordinal theory offers the following answer. During the age range when Western children first acquire meanings for written and spoken numerals, it happens that they find it much easier to learn information that is relational in nature than information that is categorical in nature. Every piece of information from the environment has both relational and categorical features. To illustrate this distinction, suppose we conduct a learning experiment in which subjects are shown pictures of pairs of dots. The members of each pair of dots are of different diameter. The subjects are taught either of two types of discriminations. First, they are taught a relational discrimination which consists of learning always to choose the larger member of each pair of dots. Second, they are taught a categorical discrimination which consists of learning always to choose a dot of a fixed size (say, 3 cm in diameter) regardless of whether it is the larger or small member of a given pair. If we find that our subjects learn the first discrimination more rapidly than the second, then we conclude that it is easier for them to learn relational information than categorical information. If we find that our subjects learn the first discrimination less rapidly than the second, then we conclude the reverse.

In the case of children in the preschool and early elementary school years, we know with reasonable certainty that relational information is less difficult to learn that categorical information. The available evidence on this matter, which is quite extensive, recently has been reviewed by P. E. Bryant (1974). Readers who are interested in the extent to which existing research confirms that subjects falling within the age range we are interested in find relational information easier to learn than categorical information are directed to Bryant's comprehensive exposition. For our part, it is sufficient to observe that if it happens to be true that preschool and elementary school children find it easier to learn relational information, then it is not surprising that, on the average, they should find it easier to learn the ordinal meanings of numerals than their cardinal meanings. As we saw in Chapter 3, ordinal numbers are purely relational constructs. When one learns the ordinal number 3, for example, one learns a certain relationship ("greater than") between 3 and prior ordinal numbers, and one learns a certain relationship ("less than") between 3 and subsequent ordinal numbers. In contrast, we saw in Chapter 4 that cardinal number is a purely categorical notion. When one learns the cardinal number 3, for example, one does not learn relations between 3 and other cardinal numbers. One learns, instead, a certain fixed manyness, which is variously called "trio" or "threeness." Hence, the forms of perceptible environmental information from which the ordinal meanings of number symbols would be derived would always be relational, whereas the forms of perceptibie environmental information from which the cardinal meanings of these same symbols would be derived would always be categorical.

In short, it seems that we can satisfactorily explain the underlying reasons for the predictions set forth above in terms of what existing research on children's learning has shown about their capacity to learn relational and categorical information. However, the reader will note that we do not yet know whether the predictions themselves are true. We do not know whether the functional relationships between ordination-cardination and arithmetic are as the ordinal theory specifies. We also do not know whether young children actually do find it easier to learn the ordinal meanings of numerals than to learn their cardinal meanings. It is the chief aim of this chapter to determine whether or not these predictions are correct. Below, we shall review two experiments, carried out by the author, which were designed to test the predictions. The first experiment is concerned with the functional connection between arithmetic and ordination-cardination, while the second experiment is concerned with how preschool children learn to associate the ordinal and cardinal meanings of the first five numerals with the symbols themselves.

EXPERIMENT 1

Although, as we noted in Chapter 6, normative studies of the sort considered in the preceding chapter suffice to establish whether there are developmental lags between ordination, cardination, and arithmetic, they do not suffice to establish the functional relationships between these variables. Just because ordination, for example, seems invariably to precede arithmetic ideas in children's thinking, it does not necessarily follow that the emergence of arithmetic ideas is somehow directly dependent on ordination. To address such a question, experimental research is necessary. That is, we must produce changes, in the laboratory, in those variables on which we believe arithmetic depends and then ascertain whether the induced changes produce corresponding changes in arithmetic. The experiment we shall now consider was designed to do just this.

Subjects

A total of 240 kindergarten children served as subjects in the experiment. These children were selected on the basis of how they performed on series of pretests which are described below. When the experiment began, the average age of the subjects was five years, two months. The experiment was conducted during the first two months of the school year before the participating children had yet received systematic instruction in arithmetic.

Method and Procedure

The experiment was conducted in three phases: pretests, training trials, and posttests. The procedures for the phases were different and, hence, will be described separately.

Pretest Phase

To begin with, the language, ordination, cardination, and arithmetic tests described in the preceding chapter were administered to kindergarten children enrolled in several elementary schools located in middle-class, residential areas of Edmonton. A total of 518 children were tested during the pretest phase. After all the tests had been administered, the children were classified according to their levels of ordination, cardination, and arithmetic proficiency. As participants for the training phase of the experiment, we were interested in identifying children who (1) were not yet capable of complete internal ordering (children who were functioning at either level I or level II of ordination), (2) were not yet capable of complete internal correspondence (level I or level II of cardination), and (3) were not yet functioning above the lowest of the three levels of arithmetic proficiency described in the preceding chapter. A total of 240 subjects (120 boys and 120 girls) who met all three of these conditions were selected for participation in the training phase. These subjects were divided into four groups of 60 children each (30 boys and 30 girls). During the training phase, the subjects in one group received special instruction in ordination; the subjects in a second group received a series of ordination problems without explicit instruction; the subjects in a third group received special instruction in cardination; and the subjects in a fourth group received a series of cardination problems without explicit instruction.

Training Phase

The training phase of the experiment spanned eight weeks. During this period, each of the 240 subjects received one training session a week for a total of eight training sessions per subject. Each of these sessions lasted for roughly 30 minutes. The procedure employed during the training sessions was different for each of the four conditions.

Ordination training. The 60 subjects in this group received a form of instruction that was designed to induce internal ordering of everyday transitive-asymmetrical relations. During each training session, the subjects were administered a graded four-step series of ordering problems involving such common transitive-asymmetrical relations as "taller than" and "larger than." In the first step of the series, the subjects were given three objects which could be ordered according to some transitive-asymmetrical relation. For example, the subjects were given three dolls that differed in height by small amounts. The correct ordering was not apparent to perception; however, it could be determined by carefully comparing the objects in pairs. The experimenter asked the children to put the three objects in order. If a child ordered them correctly, the experimenter informed the subject that the response was correct, provided a reward (candy), and went on to step two. If the three objects were not ordered correctly, the experimenter stated that the ordering was incorrect and asked the subject to try again. The subject was required to repeat the task until the correct ordering was obtained. In the second step of the series, the subjects were given five objects which could be ordered according to the same transitive-asymmetrical relation as the three objects employed on the first step. Except for the fact that five objects were employed, the second step was the same as the first; that is, the subjects were required to order the five objects as many times as was necessary to obtain the correct ordering, the experimenter informed the subjects when their responses were incorrect, and the experimenter provided a reward for correct responses. The third step of the series was the same as the first two, except the subjects were asked to order seven objects according to the same transitive-asymmetrical relation as before.

The fourth and final step was somewhat different. It resembled the ordination tests employed in the earlier normative studies. To begin with, the experimenter reintroduced the three objects employed on the first step of the present series. Next, while the subject watched, the experimenter established the transitive-asymmetrical relation that obtained between the first and second objects and the second and third objects. After these initial comparisons had been made and the subjects noted the results, the experimenter posed the ordination questions described in Chapter 8 (see pp. 134-35). Each time the subject gave a correct answer, a reward was provided. Each time the subject gave an incorrect answer, the experimenter noted the error verbally, and then explained to the subject why the answer was incorrect. After this fourth step had been completed, all four steps were repeated in order. Thus, each ordination session consisted of two administrations of a graded four-step series involving the internal ordering of objects according to some everyday transitive-asymmetrical relation. A different transitive-asymmetrical relation was employed during each of the eight sessions.

Ordination control. The 60 subjects in this group served as controls for the subjects in the first group. During each session, they were given the same series of items as the subjects in the first condition. However, they were not given any explicit instruction. On the first three steps, the objects were simply presented to the subjects and they were allowed to play with them. On the fourth step, the procedure was the same as for the subjects in the first condition, except that the experimenter did not tell the subject which answers were correct and which were incorrect. The experimenter also did not provide any reinforcements for correct judgments.

Cardination training. These subjects, like those in the first condition, received explicit instruction. However, the instruction was concerned with cardination. During each of the eight training sessions, the subjects in this condition were administered a graded four-step series of cardination problems designed to induce an understanding of the connection between relative manyness and the type of correspondence that obtains between two classes. Each of the four steps in the series consisted of six items which resembled the items on the cardination test outlined in Chapter 8. In each of the four steps, six pictures were used as stimuli. The pictures were of the same general sort as those shown in Figure 8.1. That is, each picture contained two collections of common objects lined up in parallel rows. The upper row contained more elements than the lower row in two of the pictures (as in items A and F in Figure 8.1); the upper row contained fewer elements than the lower row in two of the pictures (as in items B and D); and, in two pictures, the rows contained equally many elements (as in items C and E). The four steps in the cardination procedure, like the four steps in the ordination procedure, were graded in terms of how many objects were involved. The collections appearing in the six pictures employed in the first step contained either one, two, or three elements; those appearing in the six pictures employed in the second step contained either two, four, or six elements; and in the six pictures employed in the third step the collections contained either four, six, or eight elements. The collections appearing in the six pictures employed on the fourth step, like the stimuli for the cardination test in Chapter 8, contained either six, eight, or ten elements. Except for the absolute manyness of the collections that the subjects were shown, the training procedure for each of the four steps was the same. First, the experimenter gave the same cardination instructions as in the earlier normative studies. Next, the subject was shown one of the six pictures, and the experimenter posed the two cardination questions which we encountered in Chapter 8 (see p. 137). Whenever the subject answered correctly, a reward was given. Whenever the subject answered incorrectly, the experimenter noted the error verbally and then explained why the subject's response had been incorrect. After the subject had been shown all six of the pictures for a given step and had answered all 12 questions, the experimenter proceeded to the next step. After all four steps had been completed, the entire procedure was repeated once. Thus, each cardination-training session consisted of two administrations of four sets of stimuli graded according to how many elements the stimuli contained. In conjunction with these stimuli, the subjects were reinforced whenever they made a correct relative manyness judgment, and they were corrected whenever they made an incorrect relative manyness judgment. It should be noted that different collections of objects always were employed on successive training sessions. Thus, during one training session, for example, the subjects might make relative manyness judgments about collections of dogs and cats, whereas they might make relative manyness judgments about collections of pucks and hockey sticks during the next training session.

Cardination control. The 60 subjects in the last condition served as controls for the subjects in the cardination-training condition. During each of their eight sessions, they were given the same graded series of cardination items as the subjects in the third condition. However, the items were not accompanied by instruction. That is, correct cardination judgments were not reinforced and incorrect cardination judgments were not corrected by the experimenter. The items were simply administered to the subjects without comment in much the same manner that the cardination test was administered to the subjects who participated in the earlier normative studies.

Posttest Phase

One week after the last training session had been completed, all 240 subjects were given a series of posttests. The aim of the posttests was to determine how much progress the subjects in each of the four conditions had made in the ordination, cardination, and arithmetic spheres. However, there was special concern to determine whether or not ordination, or cardination instruction, or both, had improved arithmetic proficiency. With the exception of the language tests, the specific tests administered during the posttest phase were the same as those the subjects had taken 10 weeks earlier during the pretest phase.

Principal Findings

Ordination and Cardination

First, let us consider whether or not it was possible to produce measurable improvements in the subjects' ordination and cardination concepts as a function of the forms of instruction just outlined. Most readers no doubt expect that the aforementioned training experiments probably should have produced marked improvements in the target concepts and, consequently, may view the question of specific training effects as a somewhat trivial matter. However, the reader's reasonable, common-sense expectation of training effects is challenged by certain contemporary theories of cognitive development. For example, Piaget's theory (Piaget 1950; Piaget and Inhelder 1969), which is by far the most influential theory of cognitive development of our time, takes a dim view of attempts to induce improvements in children's logical and mathematical concepts in the laboratory. As we noted earlier on, Piaget believes that fundamental logical and mathematical concepts, especially numerical concepts, are somehow "deduced" from underlying cognitive structures. Now, it must be admitted that Piaget's cognitive structures are singularly recondite entities about which nothing positive is known. At present, about the only thing that can be said with confidence about them is that they are completely circular constructs because no procedures exist for measuring them independently of the specific concepts that are said to be deduced from them. Fortunately, we may ignore the obscurity and circularity of these structures. For our purposes, the important point about them is the manner in which they are supposed to emerge during mental development.

According to Piaget, his cognitive structures emerge "spontaneously" during the course of "natural" cognitive growth and not as a consequence of systematic instruction. The basic elements of the structures ostensibly are part of our species-specific hereditary endowment, but massive amounts of natural experience and spontaneous self-discovery are assumed to be absolutely essential preconditions for actualizing the structures (Piaget 1970b; Inhelder and Sinclair 1969; Sinclair 1973). What these claims presumably add up to is this. Since cognitive structures are partially inherited and necessitate a wide range of spontaneous experiences before they can be actualized, it is impossible to induce them via laboratory training in children who do not already possess them. This somewhat Rousseau-like hypothesis leads, in turn, to rather pessimistic predictions about the possibility of producing improvements in concepts which are believed to presuppose the structures. When a given child shows no evidence of a certain structure-related concept, the theory posits that we can usually interpret this fact to mean that the underlying structure which the concept presupposes has not yet developed. If we try to train the concept directly in the laboratory, we shall not be able to produce very significant improvements because the concept presupposes an underlying structure that is absent. If we try to train the concept indirectly in the laboratory by training the structure, the improvements also will be meager because the structures cannot be trained in subjects who have not yet acquired them spontaneously. On the whole, therefore, Piaget's theory offers a gloomy assessment of our chances of producing substantial improvements in numerical concepts such as ordination and cardination through laboratory training.

Despite the somewhat less than optimistic predictions of Piagetian theory, clear improvements were noted in both ordination and cardination, especially ordination, as a result of the training sessions. The principal findings of the experiment appear by condition and type of test in Table 9.1. To determine whether or not children's ordination concepts improved as a function of training, it is necessary to compare the average amount of pretest-to-posttest improvement for the subjects in the first condition (ordination training) with the average amount of pretest-to-posttest improvement for the subjects in the second condition (ordination control). It clearly is not sufficient simply to look for improvements in the subjects trained on ordination. Over as long a period as 10 weeks, one must anticipate spontaneous gains in all the variables measured on the pretests. The pretest-to-posttest ordination gains of the subjects in the second condition, who received the same tasks as the subjects in the first condition but were given no explicit ordination instruction, provide an estimate of how much spontaneous improvement took place. The amount of spontaneous improvement in the second condition may be used to correct the amount of improvement in the first condition to yield the true amount of training-related improvement. On the pretests, it can be seen from Table 9.1 that the subjects in the first and second conditions answered roughly 47 percent of the items correctly. On the posttests, the subjects in the first condition answered roughly 94 percent of the items correctly for a gain of 47 percent, whereas the subjects in the second condition answered roughly 55 percent of the items correctly for a gain of 8 percent. Thus, the amount of ordination improvement which the present training procedure induced in the 60 subjects in condition 1 is 39 percent, on the average.

Turning to cardination, to determine the amount of improvement in this concept as a function of training, it is necessary to compare

TABLE 9.1

Average Pretest and Posttest Scores of the Subjects Participating in Experiment 1

Type of Test	Condition			
	Ordination training	Ordination control	Cardination training	Cardination control
<u>Pretests</u>				
Ordination	5.64	5.64	5.64	5.64
Cardination	0.96	0.96	0.96	0.96
Arithmetic	5.12	5.12	5.12	5.12
<u>Posttests</u>				
Ordination	11.38	6.60	6.24	6.13
Cardination	1.32	1.19	2.76	1.56
Arithmetic	14.13	8.32	8.96	7.88

Note: High possible scores: ordination, 12; cardination, 12; arithmetic, 32.

Source: Compiled by the author.

the pretest-to-posttest gains observed in condition 3 (cardination training) with the gains observed in condition 4 (cardination control). It can be seen from Table 9.1 that the subjects in the third and fourth conditions answered roughly 8 percent of the cardination pretest items correctly. On the posttests, the subjects in the third condition answered 23 percent of the same items correctly for a total gain of 15 percent. The subjects in the fourth condition answered 13 percent of the items correctly for a total gain of 5 percent. Hence, the average net gain in cardination performance as a function of training was 10 percent.

Statistically speaking, these findings are highly significant. There is roughly one chance in five million that the difference between the ordination performance of the condition 1 subjects and the ordination performance of the condition 2 subjects was due to chance factors rather than to training. Similarly, there is roughly one chance in 500 that the difference between the cardination performance of the condition 3 subjects and the condition 4 subjects can be attributed to chance alone.

Transfer to Natural Number

We now examine whether the ordination and cardination instruction tended to transfer to the arithmetic area and, if so, whether one of the two forms of instruction tended to transfer better than the other. Both of these questions are directly relevant to the functional connections posited in the ordinal theory between ordination-cardination and how the child first learns to operate on numerals. To conduct the transfer analysis, it was once again necessary to compare the two training conditions with their respective control groups. As was the case when we considered specific training effects, it is not sufficient simply to look for pretest-to-posttest gains in the arithmetic performance of the subjects in condition 1 and condition 3. It is reasonable to suppose that arithmetic, like ordination and cardination, would undergo some spontaneous improvement during an interval of 10 weeks. The supposition is strengthened by the fact that during this same interval, all 240 subjects were receiving arithmetic instruction in their classrooms. The period during which the experiment was conducted was sufficiently long that gains in arithmetic as a function of classroom instruction undoubtedly had time to accumulate.

To determine whether or not arithmetic tended to improve as a function of ordination training, the average pretest-to-posttest improvement in the arithmetic test performance of the subjects in condition 1 was compared to the average improvement in the arithmetic test performance of the subjects in condition 2. Table 9.1 indicates that the subjects in conditions 1 and 2 answered an average of 5.12, or 16 percent, of the arithmetic test items correctly during the pretest phase. During the posttest phase, the 60 subjects trained on ordination answered an average of 14.13, or 44 percent, of the arithmetic test items correctly, whereas the ordination controls answered an average of 8.32 (26 percent) of the items correctly. Therefore, the subjects in condition 1 evidenced a total gain in arithmetic performance of 28 percent over the 10-week interval. When this value is corrected by taking into account the 12 percent gain noted in the control subjects across the same interval, we are still left with quite substantial net gain of 16 percent in the arithmetic proficiency of ordination trained subjects. Tests of statistical significance indicated that there is less than one chance in 5,000 that this net gain was due to random factors rather than the effects of ordination instruction.

To determine whether or not cardination training tended to transfer to arithmetic, the average pretest-to-posttest improvement in the arithmetic test performance of the subjects in condition 3 was compared to the average improvement in the arithmetic test performance of the subjects in condition 4. On the pretests, the subjects in these two conditions, like the subjects in condition 1 and 2, answered an average of 5.12 (16 percent) of the arithmetic items correctly. During the posttest phase, the cardination trained subjects answered an average of 8.32 (28 percent) of the arithmetic items correctly. The cardination controls answered an average of 7.88 (25 percent) of the posttest arithmetic items correctly. When the average gain in condition 3 is corrected by taking into account the average gain in condition 4, we are left with only a 3 percent improvement in arithmetic proficiency as a result of cardination instruction.

Tests of statistical significance indicated that there is an unacceptably high probability that the 3 percent difference between conditions 3 and 4 could have been due to chance rather than cardination instruction. Statistically speaking, therefore, it is impossible to conclude that cardination training produced any real improvement in arithmetic.

On the whole, the findings of experiment 1 tend to confirm the functional predictions of the ordinal theory of number development. First, there was clear evidence that gains in arithmetic proficiency accrued as a consequence of training kindergarten children to acquire internalized ordering. Second, the amount of improvement observed in arithmetic as a function of training ordination was much greater than the amount of improvement as a function of training cardination. In fact, cardination training did not produce any statistically significant improvement in arithmetic. This particular finding is not explicitly predicted by the ordinal theory. All the theory predicts is, whatever the absolute effects of ordination and cardination instruction on arithmetic, the amount of improvement which accrues from the former will be greater than the amount of improvement which accrues from the latter. Thus, experiment 1 leaves open the possibility that a more intensive program of instruction in cardination might produce some significant gains in children's understanding of arithmetic.

EXPERIMENT 2

The findings considered so far appear to indicate that, as the ordinal theory predicts, there is a closer functional relationship between understanding ordination and learning to do arithmetic than between understanding cardination and learning to do arithmetic. However, we do not yet know why this is so. Earlier on, we considered a tentative explanation proposed by the ordinal theory. It must be admitted that learning how to do arithmetic consists, at least in the beginning, of learning how to manipulate spoken or written numerals, especially the latter, in certain prescribed ways. Certainly, there can be no doubt that the arithmetic proficiency tests employed in experiment 1 and in the last two normative studies reported in Chapter

8, whatever else they may measure, measure children's ability to manipulate the first few numerals satisfactorily. It is quite reasonable to suppose that children's manipulation of these numerals will be more adequate when they can be assigned concrete meanings of some sort than when they cannot. Of course, the ordinal theory assumes that at least at the age level when arithmetic is first learned, it is considerably easier to associate numerals with their ordinal meanings than it is to associate them with their cardinal meanings. That is, for example, it is easier to learn that "2" means "second" than it is to learn that "2" means "pair." This assumption was shown to follow from what we currently know about how young children learn categorical and relational information. If the assumption happens to be correct, then it is understandable that, in experiment 1, arithmetic performance tended to improve more as a consequence of ordination training than as a consequence of cardination training. The subjects in conditions 1 and 3 both were taught concepts that, in so far as logic is concerned, may be used to give numerals concrete meaning. However, the subjects in condition 1 were taught a concept that, psychologically speaking, is intrinsically easier for them to apply to numerals than the concept which the subjects in condition 3 were taught.

Experiment 2 was explicitly designed to test the validity of the preceding argument. Before examining the details of this experiment let us consider what sort of design would be necessary to test the argument. It would seem that four general conditions must be met. First, it obviously would be necessary to work with subjects who do not yet grasp either the ordinal or cardinal meanings of "1," "2," "3," and so on. In most Western countries, where children receive systematic arithmetic instruction from age five onward, this requirement entails that the subjects must be preschoolers. Second, the subjects must be carefully pretested for their knowledge of the ordinal and cardinal meanings of numerals. Although preschool age children have not been given systematic instruction as yet, they do receive considerable unsystematic exposure to numerical ideas from other sources (for example, parents, television). Hence, it is not safe to assume that a given child does not understand the ordinal or cardinal meanings of numerals just because he or she is a preschooler. To be certain, we must pretest potential subjects and retain only those who do not evidence either meaning. Third, some of the subjects should be trained to associate numerals with their ordinal meanings and other subjects should be trained to associate numerals with their cardinal meanings. Fourth and finally, the relative ease with which these two meanings are learned must be evaluated. If it turns out that, on the average, preschoolers more rapidly learn to associate numerals with their ordinal meanings than their cardinal meanings, then the argument reviewed above would seem to be plausible. However, if the ordinal and cardinal meanings are equally difficult to learn or if the cardinal meaning is easier to learn, the argument clearly is incorrect.

Method and Procedure

Subjects

To begin with, 159 children were selected from the enrollment lists of four publicly sponsored preschools in Edmonton. All of these children were administered the pretests for ordinal and cardinal number which will be described below. A total of 120 of these subjects was retained for the training portion of the experiment. The average age of these latter subjects was four years, seven months. The oldest subject in the group was five years, five months, and the youngest subject was four years. The final sample of 120 subjects consisted of 60 boys and 60 girls.

Materials

This experiment, like the first one, was divided into three parts—pretests, training trials, and posttests. Different sets of stimulus materials were employed during each phase.

Pretest Phase. Three types of stimulus materials were employed on the pretests—numeral stimuli, ordinal-number stimuli, and cardinal-number stimuli. The numeral stimuli consisted of five 3" x 5" index cards. One of the first five numerals ("1," "2," "3," "4," and "5") appeared on each of these cards. The stimuli for ordinal number were two 8-1/2" x 11" photographs. At the top of each photograph, the first five numerals appeared in box. At the bottom of each photograph, a row of five triangles appeared. The triangles were of perceptibly different heights. In one of the two photographs, the five triangles were arranged from left to right in order of increasing height, while, in the other, they were arranged from left to right in order of decreasing height. The first of the two ordinal stimuli is shown in Figure 9.1.

The cardinal-number stimuli also consisted of two 8-1/2" x 11" photographs. The stimuli were very similar to those described for ordinal number. The first five numerals again appeared in a box at the top of the photograph. At the bottom of the photograph, a row of five boxes appeared. Within each box, there were either one, two, three, four, or five triangles. All the triangles appearing in the boxes were the same size, and they were placed at random locations within each box. Since these were cardinal-number rather than ordinal-number stimuli, the boxes containing the triangles were arranged in random orders. The order of the triangles in one photograph was three,

FIGURE 9.1

Type of Stimuli Employed in the Ordinal-Number Pretests

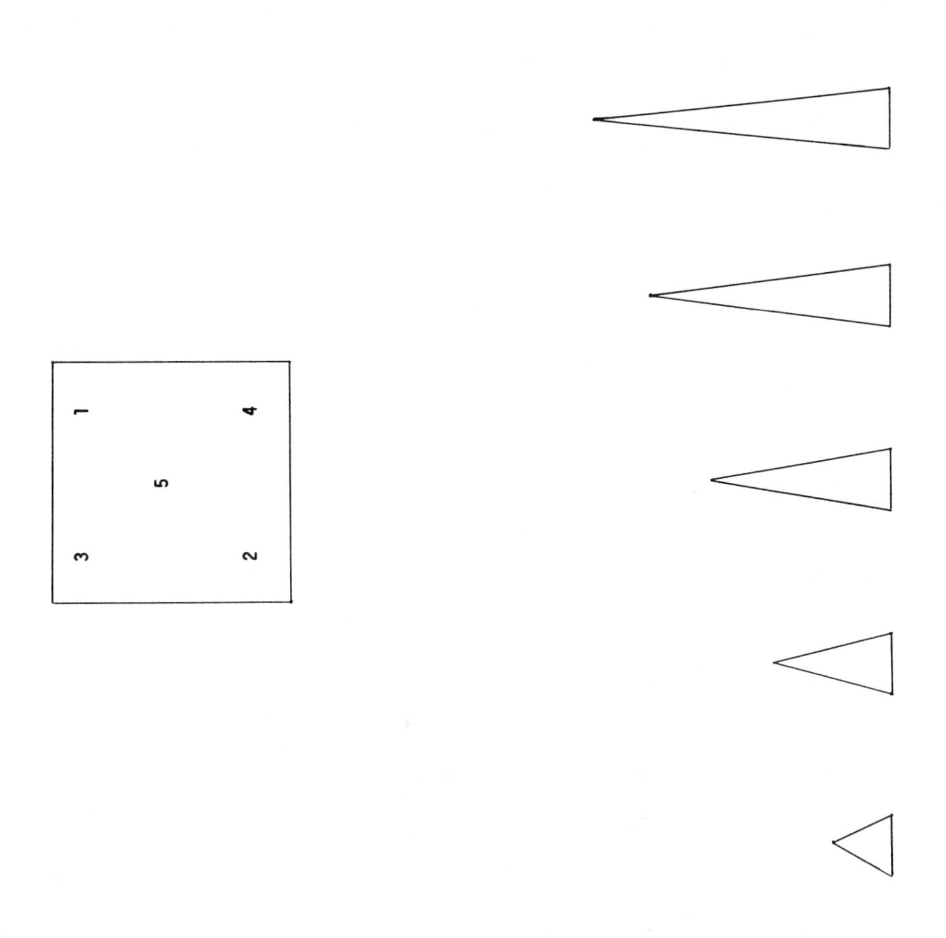

Source: Constructed by the author.

one, four, two, five. The order in the other photograph was two, four, one, five, three. The first of these two stimuli is shown in Figure 9. 2.

Training Phase. The ordinal and cardinal stimuli employed on the training trials were very similar to the pretest stimuli. There were 12 ordinal stimuli in all. Like the pretest ordinal stimuli, each training stimulus was an 8-1/2" x 11" photograph. At the top of each, a box containing the first five numerals appeared. At the bottom, a row of five simple geometric figures appeared. Each of six types of figure was displayed in two of the stimuli: circles, squares, ellipses, diamonds, and pentagons. The figures appearing on each photograph, whatever they happened to be, were always of perceptibly different heights. For each pair of photographs containing a given type of figure, the figures were arranged from left to right in order of increasing height in one photograph and from left to right in order of decreasing height in the other.

There was also a total of 12 cardinal stimuli for the training phase. Each stimulus was an 8-1/2" x 11" photograph, and, at the top of each photograph, the box with the first five numerals appeared. At the bottom of each, a row of five boxes containing between one and five simple geometric figures apiece appeared. For two stimuli each, the boxes contained circles, squares, ellipses, and pentagons. As was the case for the pretest cardinal stimuli, the figures appearing in each row of boxes were all the same size. As was also the case for the pretest cardinal stimuli, the boxes were arranged in random order. For one photograph in each pair, the ordering was five, one, four, two, and three figures. For the other member of the pair, the ordering was three, two, four, one, and five.

Posttest Phase. There were three different groups of stimuli employed on the posttests. The first group consisted of the two ordinal stimuli and the two cardinal stimuli employed on the pretests. The second group of stimuli was designed to test for transfer of training. There were two ordinal stimuli and two cardinal stimuli. The former consisted of two 8-1/2" x 11" photographs with the box containing the first five numerals appearing at the top of each photograph. At the bottom of each photograph, a row of five dogs of perceptibly different sizes appeared. On one photograph, the dogs were arranged from left to right in order of increasing size, while, on the other photograph, they were arranged from right to left in order of decreasing size. The second set of cardinal stimuli also consisted of two 8-1/2" x 11" photographs with the box of five numerals appearing at the top of each photograph. At the bottom of each photograph, a row of five boxes, each containing between one and five dogs, appeared. The dogs appearing in the boxes were all the same size, and the boxes were arranged in

FIGURE 9.2

Type of Stimuli Employed in the Cardinal-Number Pretests

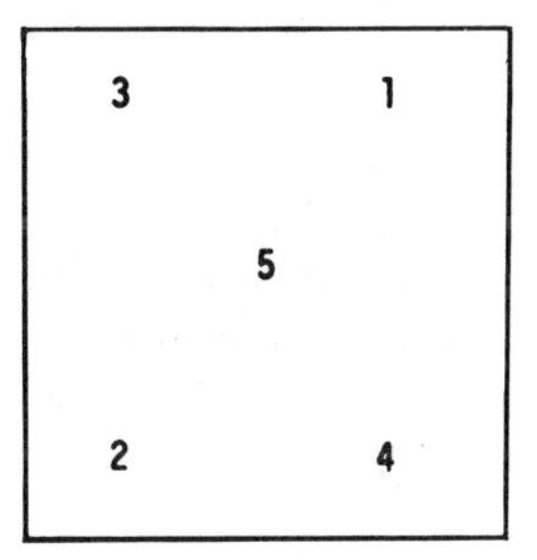

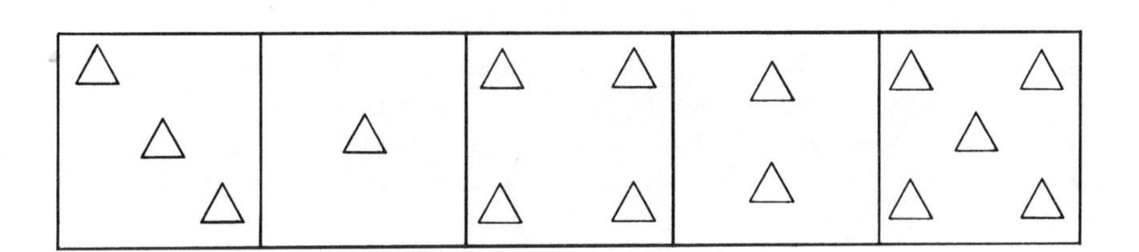

Source: Constructed by the author.

random order. The third group of stimuli also was designed to evaluate transfer. There were two ordinal-number stimuli consisting of photographs in which the box of five numerals appeared at the top and a row of five dogs of perceptibly different sizes appeared in the center. The dogs appearing in one photograph were arranged from top to bottom in order of increasing size, while the dogs in the other photograph were arranged from top to bottom in order of decreasing size. There also were two cardinal stimuli in the third group. Each stimulus was an 8-1/2" x 11" photograph in which the box containing the first five numerals appeared at the top and a row of five vertically arranged boxes appeared in the center. Each box contained between one and five dogs, and the boxes were arranged from top to bottom in random order. The dogs appearing in all of the boxes were the same size.

Testing and Training Procedures

As noted earlier, the experiment was divided into three phases. There were three different sessions spaced one week apart. The pretests and 12 training trials were administered during the first session. Four more training trials and the posttests were administered during the second session. The posttests were readministered during the third session.

Pretest phase. The 159 subjects in the original sample were pretested for their capacity to do three things: (1) identify the first five numerals, (2) associate each of the first five numerals with its correct ordinal meaning, (3) associate each of the first five numerals with its correct cardinal meaning. Only subjects who were capable of (1) but were incapable of (2) and (3) were retained for the training portion of the experiment.

The pretest for identification of the first five numerals consisted simply of showing the subjects the 3" x 5" cards mentioned earlier. The cards were presented in a random order, and the subject was asked to give the name of the numeral appearing on each card. The pretest for ordinal number consisted of showing the subject the two ordinal pretest stimuli described earlier. One of the stimuli was introduced, and the child was told that the task was to determine which of the numerals in the box at the top of the photograph went with each triangle appearing in the row at the bottom of the photograph. The experimenter then proceeded to point to each triangle in the row and ask, "Which of the numbers goes with this one?" The testing always went from the subject's left to the subject's right. Next, the second ordinal stimulus was introduced and the pretest was repeated. In the cardinal pretest, the experimenter introduced one of the two stimuli mentioned earlier and told the child that the task was to determine which of the five numerals in the box at the top of the photograph went with each of the five collections of triangles at the bottom of the photograph. Next, the experimenter pointed in turn to each collection and asked, "Which of the numbers goes with this many things?" As in the ordinal pretests, the experimenter proceeded from the subject's left to the subject's right. After the subject had been questioned about each collection in the first stimulus, the second stimulus was introduced and the cardinal pretest was repeated.

After all three pretests had been completed, the subject's answers were examined. The 120 subjects who were selected to participate in the experiment met the three general criteria: First, they were able to name all five of the numerals correctly on the numeral-identification pretest; second, they were able to choose the correct numeral on no more than four of the ten ordinal pretest items; third, they were able to choose the correct numeral on no more than four of the ten cardinal pretest items.

Training phase. Following the pretests, each of the 120 subjects who was retained for the training segment of the experiment was assigned to one of four conditions: ordinal training, ordinal control, cardinal training, cardinal control. A total of 30 subjects was assigned to each of these conditions, and each group contained the same number of boys and girls. Immediately after the pretests had been completed, a total of 12 training trials was administered to all the subjects. One week later, just before the initial group of posttests was administered, the subjects received four more training trials. The specific training procedure was different for each of the four conditions. Thus, there were 16 training trials in all.

The 30 children in the ordinal-training condition were explicitly trained to associate each of the first five numerals with its correct ordinal meaning. A simple feedback-and-correction procedure was used to instruct them. Each training trial was the same as the ordinal pretest except for two things: (1) the experimenter rewarded correct numeral choices and corrected erroneous choices, and (2) one of the 12 training stimuli described earlier was used as a stimulus. For the first 12 training trials, each of the ordinal training stimuli was used once. The first training trial began with the introduction of one of the 12 training stimuli. The experimenter again informed the subject that the task was to pick the numeral in the box that went with each of the figures appearing at the bottom of the photograph. The experimenter then pointed to the first figure in the row and asked, "Which of the numbers goes with this one?" If the subject pointed to "1" or said "one," or did both, then the experimenter said, "Yes, that is the right number." If the subject selected some other numeral, the experimenter said, "No, that is the wrong number. This one (pointing to the correct numeral) is the right one." The experimenter then pointed to the second figure in the row, posed the same question as before, and provided reinforcement or correction as needed. When the experimenter had pointed to all five figures and provided reinforcement or correction for each of the subject's choices, the first stimulus was removed and a second was introduced. The entire procedure was repeated for the second stimulus and so on until all 12 of the training stimuli had been used. One week later, at the beginning of the second session, four more training trials were administered. The procedure for each of these four trials was the same as for the first 12 trials. One of the 12 training stimuli employed during the first session was randomly selected for each of the four trials administered during the second session.

The training procedure for the 30 children assigned to the ordinal-control group was the same as for the subjects in the first condition with one major exception: On each of the trials, the experimenter neither reinforced correct numeral choices nor corrected erroneous choices. Hence, the ordinal-control procedure consisted essentially of 16 repetitions of the ordinal pretest.

The 30 children in the cardinal-training condition received instruction designed to teach them to associate each of the first five numerals with its correct manyness. The training procedure was exactly the same as for condition 1, except that the previously described cardinal-training stimuli were employed. On the first training trial of the first session, the experimenter introduced one of the 12 cardinal-training stimuli and explained that the subject's task was to pick the numeral that went with each collection of figures at the bottom of the photograph. The experimenter then pointed to the first collection and asked, "Which number goes with this many things?" If the subject pointed to the correct numeral or said its name, or did both, then the experimenter said, "Yes, that is the right number." Otherwise, the experimenter said, "No, that is the wrong number. This one (pointing to whatever numeral happened to be correct) is the right one." The experimenter then pointed to the second collection, posed the question again, and provided reinforcement or feedback as needed. When all five collections had been considered, the first cardinal stimulus was removed and a second was introduced. The procedure then was repeated for the second stimulus and so on until all 12 training stimuli had been exhausted. One week later, at the beginning of the second session, four more training trials were administered. One of the 12 training stimuli used during the first session was randomly selected for each of these four trials.

Except for the omission of reinforcement and correction, the training trials for the cardinal-control condition were the same as for cardinal training. Hence, the cardinal-control procedure, like the ordinal-control procedure, consisted of 16 repetitions of the pretest.

Posttest phase. At the end of the second session, after the subjects had received four additional training trials, three posttests were administered. The first set of posttests consisted simply of a repetition of the ordinal and cardinal pretests. The second set of posttests also consisted of a repetition of the ordinal and cardinal pretests, except that the first of the two sets of transfer stimuli described earlier were used. The third set of posttests also were the same as the ordinal and cardinal pretests, except that the second of the two sets of transfer stimuli were employed. During the third session, one week later, all three sets of posttests were administered again.

Principal Findings

The results of experiment 2 appear in Table 9. 2 where they are arranged according to condition and type of test. As was the case for experiment 1, we must consider whether or not there was any improvement in conditions 1 and 3 as a result of training before we can take up the more important question of whether children find it easier to associate numerals with their ordinal meanings than with their cardinal meanings. Taking ordinal number first, we note in Table 9. 2 that, during the first session, the subjects in condition 1 (ordinal training) answered an average of 22. 3 percent of the ordinal pretest items correctly. During the second session, these subjects gave 89. 3 percent correct answers on the first ordinal posttest, 87. 7 percent correct answers on the second ordinal posttest, and 71. 7 percent correct answers on the third ordinal posttest. Therefore, gains of 67, 65. 4, and 49. 4 percent, respectively, were registered by the condition 1 subjects on the first set of posttests. During the third session, the subjects gave 85. 3 percent correct answers on the first ordinal posttest, 80. 7 percent correct answers on the second ordinal posttest, and 51. 3 percent correct answers on the third ordinal posttest. Hence, the condition 1 gains observed on the second set of posttests were 63, 58, and 29 percent, respectively. To determine whether these gains were actually a consequence of ordinal training or were simply spontaneous improvements resulting from repeated administration of ordinal tests, it is, of course, necessary to compare each of the preceding values with the corresponding value for condition 2 (ordinal control).

During the first session, the subjects in condition 2 answered an average of 20 percent of the ordinal pretest items correctly. During the second session, they answered 34. 7, 32. 7, and 27. 3 percent of the items on the respective ordinal posttests correctly. Thus, spontaneous

TABLE 9.2

Average Numbers of Correct Numeral Choices for the Four Conditions

Test	Condition			
	Ordinal training	Ordinal control	Cardinal training	Cardinal control
<u>Pretests (Session 1)</u>				
Ordinal	2.23	2.00	2.40	2.47
Cardinal	2.37	2.13	1.83	2.17
<u>Pretests (Session 2)</u>				
First	8.93	3.47	2.17	2.33
Ordinal	1.93	2.37	6.87	2.27
Cardinal				
Second				
Ordinal	8.77	3.27	2.23	2.47
Cardinal	1.87	2.13	6.63	2.17
Third				
Ordinal	7.17	2.73	2.27	2.43
Cardinal	2.23	2.23	5.03	2.57
<u>Posttests (Session 3)</u>				
First				
Ordinal	8.53	2.93	2.27	2.47
Cardinal	2.07	1.83	5.17	2.23
Second				
Ordinal	8.07	2.53	2.33	2.53
Cardinal	2.13	2.23	5.07	2.67
Third				
Ordinal	5.13	2.63	2.47	2.17
Cardinal	2.07	2.43	3.33	2.53

<u>Note:</u> For each cell, 10 is the highest possible number of correct choices.

<u>Source:</u> Compiled by the author.

gains of 14.7, 12.7, and 7.3 percent were registered by the ordinal controls on the first set of posttests. During the third session, the ordinal controls answered 29.3, 25.3, and 26.3 percent of the items on the respective ordinal posttests correctly. Hence, the ordinal-control gains on the second set of posttests were 9.3, 5.3, and 6.3 percent, respectively. When the aforementioned gains for the condition 1 subjects on the first set of posttests are corrected for the spontaneous gains registered on the same posttests by the ordinal controls, the net gains for condition 1 are 52.3, 52.7, and 42.1 percent, respectively. When the aforementioned gains for the condition 1 subjects on the second set of posttests are corrected for the spontaneous gains registered on the same posttests by the ordinal controls, the net gains for condition 1 are 53.7, 52.7, and 23.7 percent, respectively. We may say that, on the whole, the net improvements in the capacity of condition 1 subjects to associate numerals with their respective ordinal meanings—improvements which averaged 49 percent on the first set of posttests and 43.3 percent on the second set of posttests—were quite substantial. Moreover, tests of statistical significance indicated that there is only one chance in roughly 50,000 that the net gains in the condition 1 subjects were due to random factors rather than the explicit ordinal instruction which these children received.

Table 9.2 also indicates that the instruction which the condition 3 subjects received produced perceptible improvements in their capacity to associate numerals with their cardinal meanings. During the first session, these subjects answered an average of 24 percent of the cardinal pretest items correctly. During the second session, they answered 68.7, 66.3, and 50.3 percent of the items on the respective cardinal posttests correctly. Hence, the total gains for the first set of posttests were 44.7, 42.3, and 26.3 percent, respectively. During the second session, the condition 3 subjects answered 51.7, 50.7, and 33.3 percent of the items on the respective cardinal posttests correctly. Hence, the total gains for the second set of posttests were 27.7, 26.7, and 9.3 percent, respectively. Turning to condition 4 (cardinal control), these subjects answered an average of 24.7 percent of the cardinal pretest items correctly. During the second session, they answered 23.3, 24.7, and 24.3 percent of the items on the respective cardinal posttests correctly. Therefore, the spontaneous gains for the first set of posttests were -0.7, 0.7, and 0.3 percent, respectively. During the third session, the condition 4 subjects answered 24.7, 25.3, and 21.7 percent of the cardinal posttest items correctly for spontaneous gains of 0.7, 1.3, and -2.3 percent. When the condition 3 gains observed on the first set of posttests are corrected for the spontaneous changes in the condition 4 subjects, the net gains for condition 3 are 45.5, 41.6, and 26 percent, respectively. When the condition 3 gains observed on the second set of posttests are corrected for the spontaneous changes in the condition 4 subjects, the net gains for condition 3 are 27, 25.4, and 11.6 percent, respectively. Thus, as was the case for ordinal training, the improvements observed in the cardinally trained subjects' ability to associate numerals with their cardinal meanings—improvements which averaged 37.7 percent on the first set of posttests and averaged 21.3 percent on the second set of posttests—were substantial. Tests of statistical significance indicated that there is less than one chance in 1,000 that the net gains in condition 3 subjects' cardinal-number performance was due to random factors rather than the cardinal instruction which these subjects received.

We come now to the considerably more important question of the relative difficulty of learning to associate numerals with their ordinal meanings and learning to associate numerals with their cardinal meanings. To answer this question, it is necessary to compare the performance of the condition 1 subjects on the three ordinal posttests with the performance of the condition 3 subjects on the three cardinal posttests. Taking the second session first, it will be recalled that the net condition 1 gains on the three ordinal posttests were 52.3, 52.7, and 42.1 percent while the condition 3 gains on the three cardinal posttests were 45.4, 41.6, and 26 percent. On the average, therefore, there was 11.3 percent more improvement in the condition 1 subjects' ability to associate numerals with their ordinal meanings than in the condition 3 subjects' ability to associate numerals with their cardinal meanings. A test of statistical significance indicated that there is only one chance in roughly 5,000 that this difference between conditions 1 and 3 was due to random factors rather than to the greater intrinsic ease of learning ordinal associations. The difference between conditions 1 and 3 was roughly twice as large on the second set of posttests. During the second session, the net condition 1 gains on the three ordinal posttests were 53.7, 52.7, and 23.7 percent while the net condition 3 gains on the three cardinal posttests were 27, 25.4, and 11.6 percent. Hence, on the average, there was 22 percent more improvement in the condition 1 subjects' ability to associate numerals with their ordinal meanings than in the condition 3 subjects' ability to associate numerals with their cardinal meanings. A test of statistical significance indicated that there is only one chance in 10,000 that this difference was due to random factors rather than to the greater intrinsic ease of learning ordinal associations.

To summarize, the findings of experiment 2 suggest two general conclusions. First, it is possible, using rather simple procedures, to teach preschoolers to associate the symbols "1," "2," "3," "4," and "5" with their respective ordinal and cardinal meanings. Table 9.2 indicates that on the first posttest, condition 1 subjects, on the average, knew the ordinal meanings of between four and five of these numerals. Table 9.2 also indicates that, on the same posttest, condition 3 subjects, on the average, knew the cardinal meanings of between two and three of these numerals. On the pretests, these same subjects knew the ordinal and cardinal meanings of one numeral, on the average. Second, preschoolers find it considerably easier to learn the ordinal meanings of the first five numerals than to learn their corresponding cardinal meanings. This latter finding is of great importance because it permits us to explain the experiment 1 finding that there is a much closer connection between understanding ordination and learning arithmetic. In its initial stages. learning arithmetic consists mainly of learning how to manipulate numerals properly. Children apparently find it much easier to learn the correct ordinal meanings of these symbols than to learn their corresponding cardinal meanings.

10

THE GROWTH OF CARDINATION

We shall now take up another large-scale normative study. This particular investigation differs from the normative studies reported in Chapter 8 in that it is concerned exclusively with the development of cardination and some related cardinal ideas during the elementary school years. The study was conducted by the author during early 1974 in collaboration with M. E. Fraser and was undertaken for two principal reasons. First, it appeared desirable to have more exact information about the age at which North American children may, on the average, be expected to understand thoroughly the connection between type of correspondence and relative manyness (that is, the age at which they attain level III of cardination). In view of the restricted age ranges employed in the three studies reported in Chapter 8 (five- to six-year-olds in two studies and five- and eight-year-olds in the other study), it is impossible to be very exact about how much cardination changes with age during the elementary school years. To say with some accuracy when most children have attained level III of cardination and to say how their grasp of cardination changes with age, it is necessary to investigate the entire elementary school age range.

The reader might reasonably ask why we should wish to determine when, on the average, children have attained level III of cardination. After all, this would not seem to be a very important issue from the standpoint of the three theories of number development discussed in Chapter 6. As far as these theories' predictions about cardination are concerned, it suffices to determine the developmental and functional relationships between cardination, on the one hand, and ordination and arithmetic, on the other. Since cardination age norms apparently are of no theoretical interest, why gather data on them? The

---

This chapter was prepared with the collaboration of M. E. Fraser.

motives of the research were entirely pragmatic; although such age norms are theoretically uninteresting, their practical significance, when viewed from the perspective of contemporary North American arithmetic instruction, is very great. For some years now (fifteen, to be precise), arithmetic instruction in North American elementary schools has relied heavily on cardinal ideas. Most standardized arithmetic curricula currently in use initiate arithmetic instruction with the concepts of class, manyness, and correspondence. It is assumed that, by the time they enter elementary school, children have an intuitive grasp of the connection between correspondence and relative manyness. One of the first steps in most arithmetic curricula is to provide instruction designed to amplify and clarify children's intuitive understanding. A subsequent step is to teach children to associate numerals with their respective manynesses. Finally, arithmetic instruction (teaching children to manipulate numerals in certain prescribed ways) begins in earnest. Now, the heavy reliance on cardinal ideas as a basis for teaching elementary school children arithmetic may seem somewhat paradoxical in view of the empirical findings considered in the preceding two chapters; this paradoxical state of affairs will be considered in Chapter 11. For the present, however, we might observe that the heavy reliance on cardination as a basis for arithmetic instruction is an inescapable fact of life in modern elementary schools. Hence, if we propose to continue teaching cardination to our children (by no means is it suggested here that we should), it would be well to know exactly when this concept tends to appear in most children's thinking. We may then, at least, gear the timing of our cardination instruction to coincide with the age at which children actually possess the concept.

Our second reason for conducting this study was rather more theoretical. The ordinal theory, it will be recalled, posits that children become capable of complete internal correspondence (cardination) sometime after they become capable of complete internal ordering (ordination) and after they have made substantial progress with arithmetic. Given what we already know about when subjects are capable of internal ordering and when they begin to make progress with arithmetic, this assumption relegates the emergence of complete internal correspondence to the middle or late elementary school years, or perhaps even later. However, the assumption also leads to some ancillary predictions about the emergence of other cardinal ideas which are related to internal correspondence. Some of these ideas would seem to be necessary preconditions for cardination and, hence, we would expect them to appear in children's thinking before cardination. Other ideas would seem to presuppose cardination as a necessary precondition and, hence, we would expect them to emerge sometime after cardination. A general prediction about these cardination-related concepts which follows from the ordinal theory: During the elementary school years, we may anticipate a protracted period of development for cardination-related concepts. That is, in contrast with concepts in the ordination sphere, average children do not begin elementary school with an extensive range of cardinal skills at their disposal. Instead, they possess only the most rudimentary and percept-like cardinal notions at age five. During the elementary school period, increasingly sophisticated cardinal ideas are gradually worked out and, eventually, cardination itself is understood. After cardination has been attained, the children move on, during the juvenile and adolescent years, to even more difficult cardinal ideas which presuppose a thorough grasp of the relationship between correspondence and relative manyness.

Thus, a slow and protracted period of development for cardinal number skills, spanning at least the elementary school years, is envisioned in the ordinal theory. This prediction of the ordinal theory may be contrasted with what the cardinal and cardinal-ordinal theories would lead one to expect. The most reasonable prediction from the cardinal theory would be that children have an extensive range of cardinal-number skills at their command by the time they enter elementary school. Thus, one would expect very few age-related changes in such concepts during the elementary school years. The most reasonable prediction from Piaget's cardinal-ordinal theory is a period of cardinal-number development which is somewhat longer than what the cardinal theory predicts but considerably shorter than what the ordinal theory predicts. On the one hand, Piaget does not predict that most children will understand most cardinal concepts by age five (Inhelder and Piaget 1964). However, he does expect that most cardinal concepts will be present by age seven or eight. The latter ages are the nominal bounds for the onset of Piaget's "concrete-operational" stage of mental development, with which he associates cardinal concepts. The data on cardination reviewed in Chapter 8 indicated that, as the ordinal theory predicts, complete internal correspondence seems to emerge considerably later than either age five or age seven to eight. But from this fact it does not necessarily follow that other cardinal ideas emerge slowly throughout the elementary school years. It might well be that there is something unusual about cardination and that other related ideas appear either by age five or by age seven to eight.

To determine whether there is, in fact, a protracted period of development for cardinal skills spanning the elementary and juvenile years, five different cardinal skills were selected and measured, along with cardination, in a very large sample of children whose ages ranged from roughly four and one-half to 13 years. On the basis of past research (Brainerd and Fraser 1975; Brainerd and Kaszor 1974; Kofsky 1966), four specific ideas were believed to be necessary preconditions for complete internal correspondence: (1) the concept of class, (2) the concept of class extension, (3) the concept of cardinal equivalence, and (4) the concept of cross-classification by manyness. Each of these concepts appears to require skills which are necessary but by no means sufficient for understanding the connection between type of correspondence and relative manyness. Concept 1 refers simply to the child's capacity to understand that, for any given assortment of concrete objects, different objects belong to different and mutually exclusive classes. This is an extremely rudimentary cardinal skill which does not presuppose any knowledge of either correspondence or manyness. It was measured by a simple object-sorting task. Concept 2 refers to children's understanding that discrete collections may be classified together on the basis of the logical property of similarity (containing equally many terms) as well as on the basis of more perceptible qualities such as color or shape. This concept was measured by a card-sorting task. Concept 3 refers to children's understanding that if one-to-one perceptual correspondence is established between two discrete collections, the collections will always contain equally many terms as long as no terms are added to or subtracted from either collection. This concept was measured by establishing a perceptual one-to-one correspondence between two collections and then making the correspondence no longer perceptually apparent by rearranging the terms in one of the collections. Concept 4 refers to children's understanding that manyness can be combined with other intensions (such as color or shape) to yield cross-classifications of discrete collections of objects. This concept was measured by a matrix-completion task.

Each of the four concepts just described were believed to be logically simpler than cardination. Unlike the cardination tests described in Chapter 8, none of these concepts requires an internal grasp of the connection between type of correspondence and relative manyness. The concept of class, as we saw, is not directly concerned with either correspondence or manyness. Two of the concepts, class extension and cross-classification by manyness, are concerned with manyness but not with correspondence. Class extension requires only that children pay attention to manyness and use it as a property in virtue of which different collections can be grouped together. Cross-classification by manyness requires only that children pay attention to manyness and combine it with other properties as a basis for grouping different collections together. The concept of cardinal equivalence is the only one of the four that seems to be concerned—but only in a very limited way—with both correspondence and relative manyness. Cardinal equivalence requires, first, that children understand that a one-to-one perceptual correspondence between the terms of two discrete collections implies that the collections contain equally many terms and, second, that children understand that subsequent perceptual alterations of one of the collections does not affect the manyness relationship. This particular concept obviously is a far cry from the notion of complete internal correspondence. In view of the differences between the demands of these concepts, on one hand, and the demands of the concept of cardination, on the other, it was anticipated that most children would tend to acquire all of the former concepts before they acquired cardination.

Finally, a fifth concept was included in this study for which, it was believed, cardination is a necessary precondition. This fifth concept is the principle of class inclusion, which we have previously encountered in Chapters 6 and 7. It will be recalled that class inclusion refers to children's understanding that, given two discrete collections, A and A', which may be united to yield some superordinate class B, the superordinate class contains more terms than either of its two constituent classes. It also will be recalled that Piaget uses class inclusion as his chief index of the presence of cardination in children's thinking (Beth and Piaget 1966; Inhelder and Piaget 1964). This fact suggests that children should grasp cardination and class inclusion at about the same time. However, common sense suggests that understanding that the type of correspondence obtaining between two collections automatically determines their relative manyness is a logically simpler concept than class inclusion. In the cardination tests described in Chapter 8, the subjects must do essentially two things. First, they must spontaneously establish a term-for-term correspondence between two perceptually distinct collections and, second, they must deduce from this correspondence the manyness relationship which obtains between the collections. Class-inclusion problems require that subjects do these two things and, in addition, they also require that children mentally separate the two classes whose relative manyness is being judged. Unlike the two classes employed in the cardination tests (red dots and blue dots), the two classes with which class-inclusion problems are concerned never are physically distinct. One of them is a proper part of the other. Therefore, it is very possible that children who actually are capable of complete internal correspondence (and, hence, would pass the cardination tests) would not be capable of the additional (and irrelevant) skill demanded by class-inclusion problems and would fail such problems. For these reasons, it appears obvious that employing the solution of class-inclusion problems as one's supreme criterion of cardination, as Piaget does, is an error. Given the demands that class-inclusion problems make in addition to understanding the connection between correspondence and relative manyness, it seems reasonable to conclude that solution of class-inclusion problems may be viewed as a sufficient but not necessary criterion for the presence of complete internal correspondence. For this specific reason, it was expected that most children would not demonstrate an understanding of class inclusion problems until some time after they had attained level III of cardination.

METHOD AND PROCEDURE

Subjects

A total of 350 children served as subjects in this study. The subjects were selected randomly from the class lists of several elementary schools located in middle-class residential areas of Edmonton. There were 50 subjects (25 boys and 25 girls) from each of the grade levels from kindergarten through sixth grade. The average ages of the subjects were as follow: kindergarteners, five years, two months; first graders, six years, three months; second graders, seven years, two months; third graders, eight years, one month; fourth graders, nine years, four months; fifth graders, ten years, two months; sixth graders, eleven years, five months. The youngest subject participating in the study was four years, five months and the oldest subject participating in the study was thirteen years, one month.

Testing Procedures

All 50 subjects at each of the seven age levels were administered six different types of tests, described below. The six tests always were administered in a random order. This was done to avoid the possibility that observed differences in the ages at which subjects tend to pass the tests might be attributable to extraneous order effects such as warm-up or fatigue.

The Concept of Class

To determine whether or not a given subject understood the concept of class, he or she was administered a simple object-sorting task. First, the experimenter placed an assortment of eight small plastic chips in front of the subject. There were two square yellow chips, two square black chips, two triangular yellow chips, and two black triangular chips. Two 8-1/2" x 11" pieces of cardboard were also placed in front of the subject. The experimenter then explained that they were going to play a game in which the subject's task was to sort the colored chips into two piles, one pile on each cardboard sheet "so that all the chips in each pile are the same in some way." The experimenter also told the subject that there was no "right" or "wrong" way to play the game and that the subject should "just put the things together that go together." Of course, there were two intensions which could be used to partition the eight chips into two classes—color and shape. Next, the subject was given three minutes to partition the chips into two classes. (Pilot research had shown that most subjects who solve this task require less than one minute to do so.) When the subjects had sorted the chips into two classes, the experimenter removed the chips from the cardboard sheets, mixed them up, and placed them in front of the subject again. The experimenter then initiated a second trial by saying, "Now could you sort the chips into two piles again but this time in a different way? Remember that all the chips in each pile have to be the same in some way." Again, the subject was allowed three minutes to partition the chips into two classes. To make two consistent and exhaustive classifications of the chips, therefore, it was necessary for the subject to partition them in terms of their color on one trial and to partition them in terms of their shape on the other trial.

Class Extension

To determine whether or not a given subject understood the concept of class extension, a card-sorting task was administered. A total of eight cards were used in the task and are shown in Figure 10.1. The cards were standard 3" x 5" index cards. Note that they varied along three intensions: color (black or yellow), shape (square or triangle), and manyness (6 terms or 8 terms). The class-extension test consisted of three trials. The first trial began with the experimenter placing two 8-1/2" x 11" cardboard sheets in front of the subject. The experimenter then shuffled the eight cards and spread them out, face up, in front of the subject. The experimenter then told the subject, "We play this game with these small cards. Each card has a different picture on it. I should like you to sort the cards into two piles, one here and one here (pointing to the two cardboard sheets), so that all the cards in each pile are the same in some way. There are lots of different ways to do it, so just do it the first way you think of." The subject then was allowed three minutes in which to partition the cards into two classes. After the sorting had been completed (or after three minutes), the experimenter removed the cards from the cardboard sheets, shuffled them, and again spread them out face up in front of the subject. The second trial of the class extension test then began by the experimenter saying, "Now could you sort the cards again so that all the cards in each pile are the same in a different way?" The subject was allowed another three minutes to sort the cards a second time. The experimenter then removed the cards from the cardboard sheets, shuffled them, and spread them face up in front of the subject for a third time. The third trial began by the experimenter saying, "Now, can you sort the cards in still another way? Remember that

FIGURE 10.1

Eight Cards Employed in the Class-Extension Test

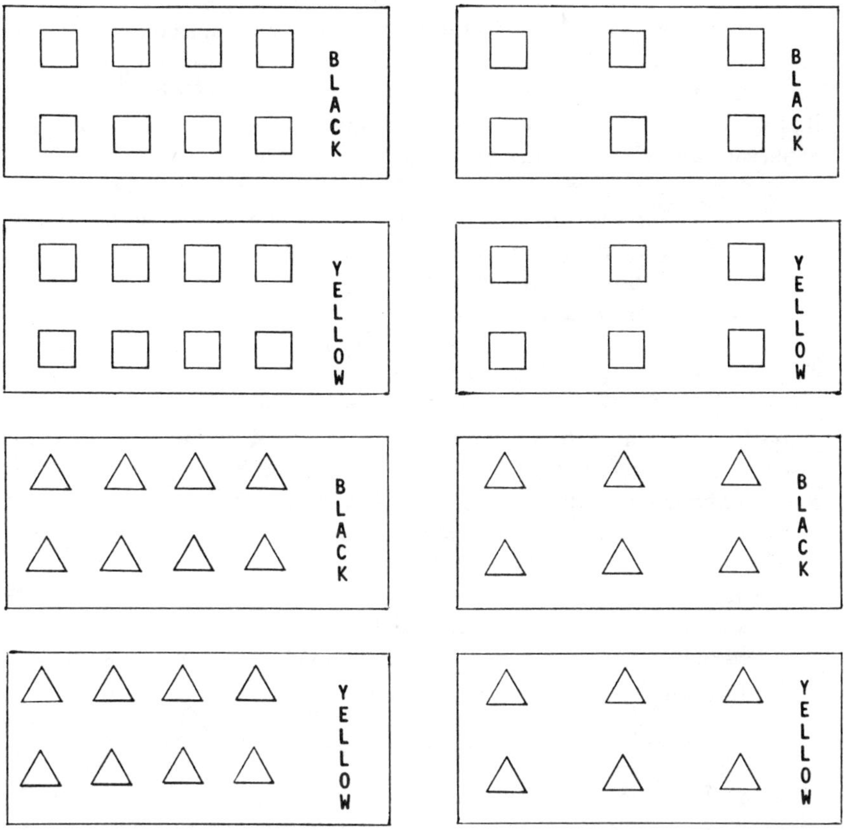

BLACK BLACK

YELLOW YELLOW

BLACK BLACK

YELLOW YELLOW

Source: Constructed by the author.

all the cards in each pile have to be the same in some way." Thus, in brief, the subjects were given three chances to sort eight cards varying along three intensions in three different ways.

Cardinal Equivalence

To determine whether or not a given subject understood cardinal equivalence, a test was given involving two rows of small plastic chips of the type employed on the concept-of-class tests. First, the experimenter placed a pile of 12 chips in front of the subject (six yellow chips and six black chips). The experimenter then introduced a second, identical pile of chips. The experimenter withdrew eight chips and constructed a row with about one inch between successive chips. The experimenter then asked the subject to construct a row of his or her own that contains "just as many chips as my row." A subject who could not construct a row containing eight chips by placing one chip across from each chip in the experimenter's row did not proceed any further in the cardinal-equivalence test. If the subject did construct a correct row, the experimenter posed three questions: (1) If I pushed the chips in my row close together so that they touched, would there still be just as many chips in my row as in yours?" (2) If I pushed the chips in my row close together so that they touched, would your row then have more chips than mine?" (3) If I pushed the chips in my row close together so that they touched, would my row then have fewer chips than yours?" After the subject had answered these three questions, the experimenter compressed the row until the chips were touching. Next, three more questions were posed: (4) Does my row still have as many chips as your row?" (5) Does your row have more chips than mine now?" (6) Does my row have fewer chips than yours now?" Thus, the cardinal-equivalence test assessed the extent to which children understood that the destruction of an initial one-to-one correspondence that they had established does not change the manyness relationship between two collections.

Cross-Classification by Manyness

To determine whether or not a given subject could combine the property of manyness with other intensions to cross-classify different collections of objects, a series of matrix-completion tasks was administered. A total of six 2 x 2 matrices were used. An illustrative matrix is shown in Figure 10. 2. The reader will note that the columns of the illustrative matrix differ in terms of manyness, differ in terms of shape, and that one cell of the matrix is empty. Below the matrix, a row of five cells appear. One of them (second from the left) is the missing cell of the matrix. The subject's task is to determine which of the five cells at the bottom is the missing cell. There were five

FIGURE 10. 2

Sample Matrix Problem Employed to Test Children's Understanding of Cross-Classification by Manyness

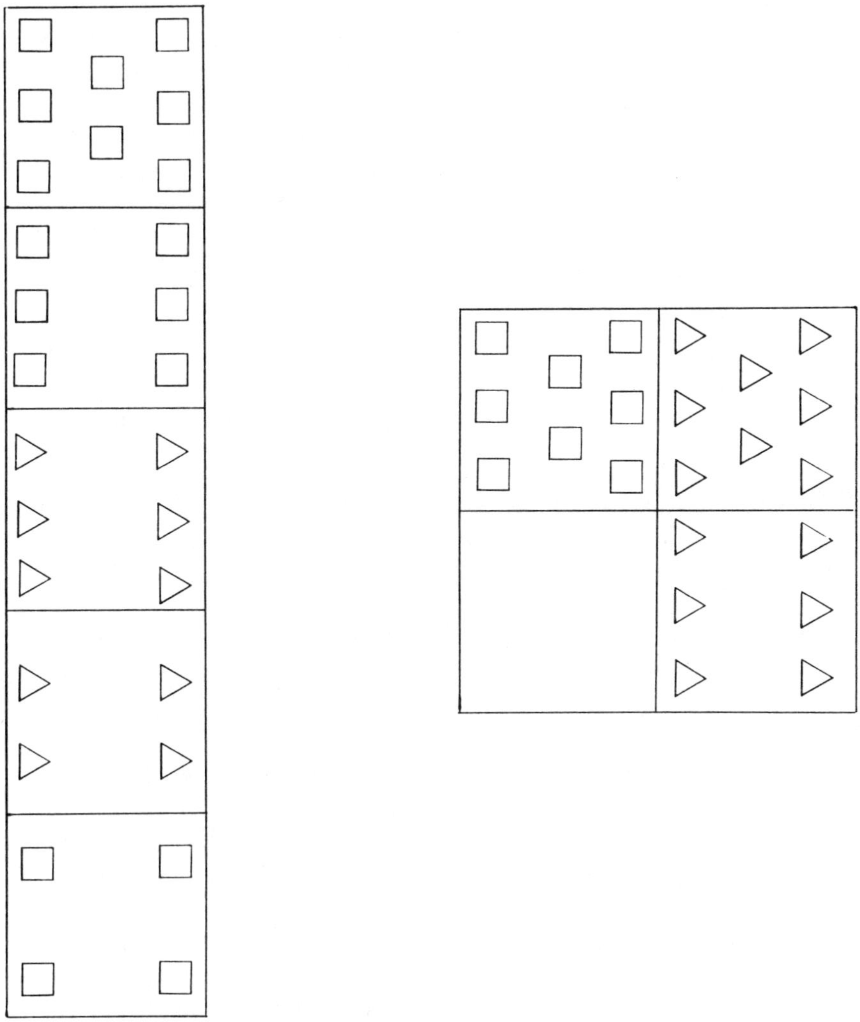

Source: Constructed by the author.

other matrixes similar to the one in Figure 10. 2. One cell was empty in all of these other five matrixes. In three of the other matrixes, the rows differed in terms of manyness (six or eight terms). In the remaining two, the columns differed in terms of manyness (six or eight terms). In three of the other matrixes, either the rows or columns differed in terms of color (black terms or yellow terms). In the remaining two, either the rows or columns differed in terms of their shape (square or triangle). The test of cross-classification by manyness consisted of six trials. At the beginning of the test, the experimenter explained that, "At the top of each card there is a box containing four pictures, but one of the four pictures in the box will always be missing. At the bottom of each card are some more pictures and one of them is the missing picture that belongs in the box." The subjects were told that there was always a "rule" by which they could figure out which of the pictures at the bottom was the missing cell. To illustrate these instructions, the experimenter introduced a practice four-cell matrix in which the columns differed in terms of shape (square or triangular terms) and the rows differed in terms of color (black or yellow terms). Each subject was asked to pick the missing cell and then to explain the choice. The experimenter then offered a complete explanation and informed the subject that there would be a similar rule by which the missing cells in the other matrixes could be filled in. The subject was asked to remember that "the rule always has two parts. One part says how the pictures are the same and the other part says how they are different." Next, the cross-classification test itself was administered. On each trial, one of the six matrixes described above was presented and the subject was asked to pick out the missing matrix cell from among the five choices at the bottom of the stimulus. After the subject had made a choice, the stimulus was removed and another matrix was introduced.

Cardination

To determine whether or not a given subject understood the connection between correspondence and relative manyness, the cardination test reported in Chapter 8 was administered to all of the subjects participating in this study.

Class Inclusion

To determine whether or not a given subject understood the class-inclusion principle, tests concerned with the relative manyness of superordinate classes and subordinate classes were administered. The class-inclusion test consisted of two trials on which different stimuli were used. The two stimuli are shown in Figure 10. 3. In the stimulus at the top of Figure 10. 3, the two subordinate classes

FIGURE 10.3

Stimuli Employed to Test Children's Understanding of the Class-Inclusion Principle

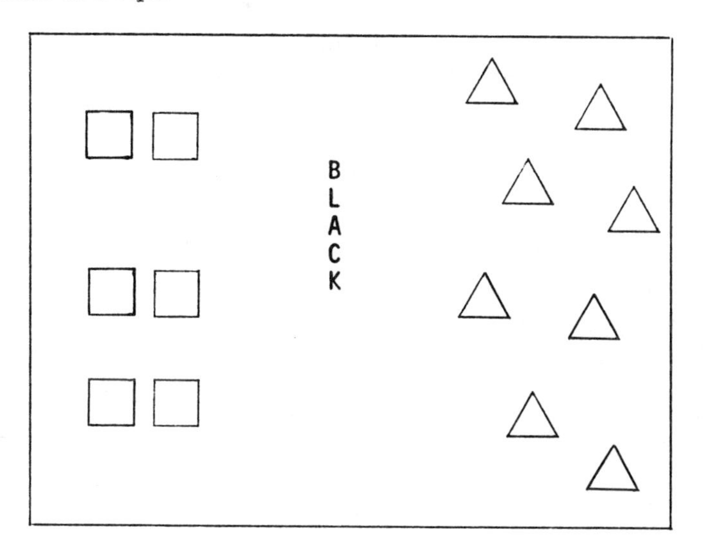

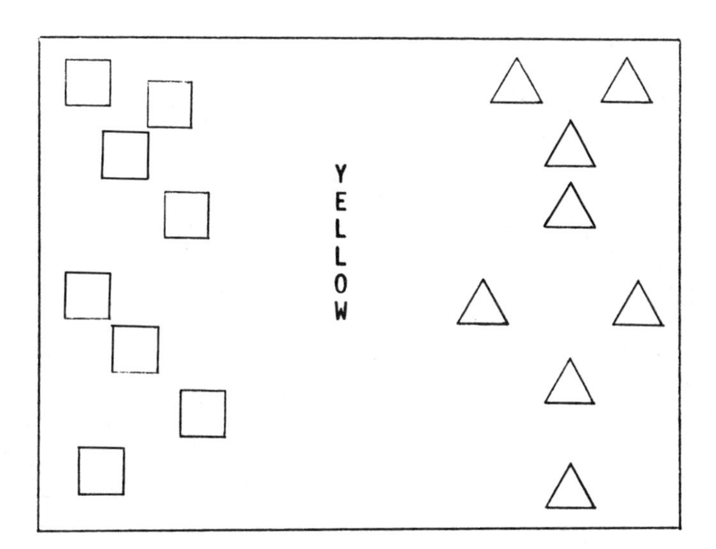

Source: Constructed by the author.

(squares and triangles) contain unequally many terms (six or eight) and the superordinate class ("black things") contains 14 terms. In the stimulus at the bottom of Figure 10. 3, the two subordinate classes (squares and triangles) contain equally-many terms (eight) and the superordinate class ("yellow things") contains 16 terms. At the beginning of the first trial, the experimenter placed one of the two stimuli in front of the subject and said, "There are several (yellow or black) things in this picture. Some of the things are triangles and some of them are squares. But all of them are (yellow or black) things. Right?" After answering affirmatively, the subject was asked to "count all the squares," then to "count all the triangles," and finally to "count all the (yellow or black) things." After the subject had counted the elements in both subordinate classes and in the superordinate class, the experimenter posed three questions: (1) Are there the same number of squares as there are (yellow or black) things? (2) Are there the same number of triangles as there are (yellow or black) things? (3) Are there more (yellow or black) things than there are squares? After all three questions had been answered, the first stimulus was removed. The remaining stimulus then was introduced and the subject was asked to "count all the squares," to "count all the triangles," and finally to "count all the (yellow or black) things." The three questions then were repeated.

PRINCIPAL FINDINGS

In view of the fact that a convenient three-level classification scheme already existed for one of the six tests administered in this study (cardination), it was decided to employ a three-level classification scheme with the other five tests also. Since the subjects were required to make different numbers of responses on different tests, the implementation of a single three-level scheme for all of them greatly facilitated the comparison of the subjects' performance. The following criteria were employed to assign each subject to one and only one level on each test:

On the class-concept test, subjects were assigned to level I if they sorted incorrectly on both trials, to level II if they sorted incorrectly on one trial and correctly on the other, and to level III if they sorted correctly on both trials. On the class-extension test, subjects were assigned to level I if they failed to group the cards according to manyness on all three trials, to level II if they grouped the cards according to manyness on the last trial only, and to level III if they grouped the cards according to manyness on either of the first two trials. On the cardinal-equivalence test, subjects were assigned to level I if they either failed to construct a row containing equally many chips as the experimenter's row or, if having done so, failed to answer all of the prediction questions correctly. Subjects were assigned to level II if they answered all three prediction questions correctly but failed to answer all three posttransformation questions correctly, and to level III if they answered all three prediction questions and all three posttransformation questions correctly. On the cross-classification tests, unlike the previous tests, there was a built-in chance solution rate (0.20). That is, on each individual trial, there was always one chance in five that the subject would respond correctly. Since there were six trials in all, we would expect either one or two correct responses for each subject by chance alone. Hence, subjects responding correctly on zero, one, or two trials were assigned to level I. Subjects were assigned to level II if they responded correctly on either three or four trials and to level III if they responded correctly on either five or six trials. On the cardination test, the same criteria employed in Chapter 8 also were used to assign the present subjects to levels of cardination. On the class-inclusion test, subjects were assigned to level I if they failed to answer all three questions correctly for at least one of the two stimuli, to level II if they answered all three questions correctly for one of the stimuli but not for the other, and to level III if they answered all three questions correctly for both stimuli.

Age-Related Changes in Cardinal Concepts

The initial question to be examined concerns whether and how children's grasp of the six cardinal concepts changes with age during the elementary school years. To examine this question, the subjects from each age level were grouped in terms of the numbers of subjects functioning at each of the three levels of the six concepts. The data for the entire sample appear in Table 10.1. To determine whether or not each concept tended to improve with age, the mean levels of each concept were compared across age levels using simple analyses of variance as tests of statistical significance. The analysis for age changes in the class concept indicated that this particular concept did not improve with age. As can be seen in Table 10.1, virtually all of the youngest subjects in the sample were functioning at level III. There were only 16 subjects (eight kindergarteners and eight first graders) in the entire sample of 350 who did not respond perfectly on the class-concept tests. The analysis for age changes in cardinal equivalence indicated that this concept improves with age during the first half of the elementary school age range. Two-thirds of the kindergarteners were functioning below level III. However, performance on the cardinal-equivalence tests was essentially perfect from grade three onward. The analyses for age changes in class extension and cross-

TABLE 10.1

Numbers of Subjects Functioning at the Three Levels for Each of the Six Concepts

Concept	Level I	Level II	Level III
		Kindergarten	
Class	4	4	42
Class extension	31	15	4
Cardinal equivalence	22	8	20
Cross-classification	41	9	0
Cardination	48	2	0
Class inclusion	48	2	0
		First grade	
Class	0	8	42
Class extension	11	21	18
Cardinal equivalence	10	15	25
Cross-classification	32	18	0
Cardination	34	16	0
Class inclusion	47	3	0
		Second grade	
Class	0	0	50
Class extension	11	21	18
Cardinal equivalence	6	10	34
Cross-classification	20	26	4
Cardination	25	22	3
Class inclusion	42	8	0
		Third grade	
Class	0	0	50
Class extension	10	10	30
Cardinal equivalence	0	9	41
Cross-classification	9	31	10
Cardination	9	34	7
Class inclusion	20	29	1

Concept	Level I	Level II	Level III
		Fourth grade	
Class	0	0	50
Class extension	6	10	34
Cardinal equivalence	0	0	50
Cross-classification	3	27	20
Cardination	8	28	14
Class inclusion	20	25	5
		Fifth grade	
Class	0	0	50
Class extension	0	15	35
Cardinal equivalence	0	0	50
Cross-classification	0	19	31
Cardination	6	27	17
Class inclusion	19	16	15
		Sixth grade	
Class	0	0	50
Class extension	0	14	36
Cardinal equivalence	0	0	50
Cross-classification	0	9	41
Cardination	3	23	24
Class inclusion	16	17	17

Source: Compiled by the author.

classification by manyness indicated that both of these concepts improved gradually throughout the entire elementary school age range. Concerning the former, roughly two-thirds of the kindergarten subjects showed no evidence of class extension. During the subsequent five years, the concept improves steadily until, by age 10, 70 percent of the children are functioning at level III. Similarly, roughly 80 percent of the kindergarten subjects showed no evidence of cross-classification by manyness. During the subsequent six years, the concept improves gradually until by age 11, roughly 80 percent of the subjects have attained level III.

Cardination and class inclusion clearly evolve more slowly than the preceding four concepts. In fact, cardination and class inclusion evidenced a pattern of age changes that is essentially the reverse of the one described above for cardinal equivalence. That is, there was no improvement during the first few years followed by gradual improvement during the later years. Taking cardination first, the great preponderance of kindergarteners and first graders were functioning at level I. Moreover, it was not until age 9 that an appreciable number of children (14) evidenced complete internal correspondence. Even by age 11, slightly less than 50 percent of the children evidenced complete internal correspondence. Thus, it seems that it is not until sometime during adolescence that we may say that most subjects will have attained level III of cardination. Turning to class inclusion, children appear to make absolutely no progress with this concept before roughly age eight. Between age eight and age 11, the proportion of subjects who clearly understand that a superordinate class always contains more elements than any one of its subordinate classes gradually increases. However, the amount of improvement in class inclusion during this four year period could hardly be called dramatic. By age 11, only about one subject in three had attained level III. Interestingly, roughly the same proportion of 11-year-olds are still functioning at level I.

The patterns of age-related variations just considered are shown graphically in Figure 10.4. It is easy to see that the concept of class is the only one of the six cardinal skills of which it may safely be said that virtually all children are capable of it when they enter elementary school. It is extremely interesting to observe that this also is the only one of the six concepts which does not involve either manyness or correspondence. It can also be seen from Figure 10.4 that children enter elementary school with some slight understanding of cardinal equivalence and class extension. However, they show no evidence of cross-classification by manyness, cardination, and class inclusion until much later. The latter two concepts are never understood by more than a minority of elementary school children. Of course, cardination and class inclusion are the only concepts of the six which require that the

FIGURE 10.4

Average Age-Related Improvements in Cardinal Concepts

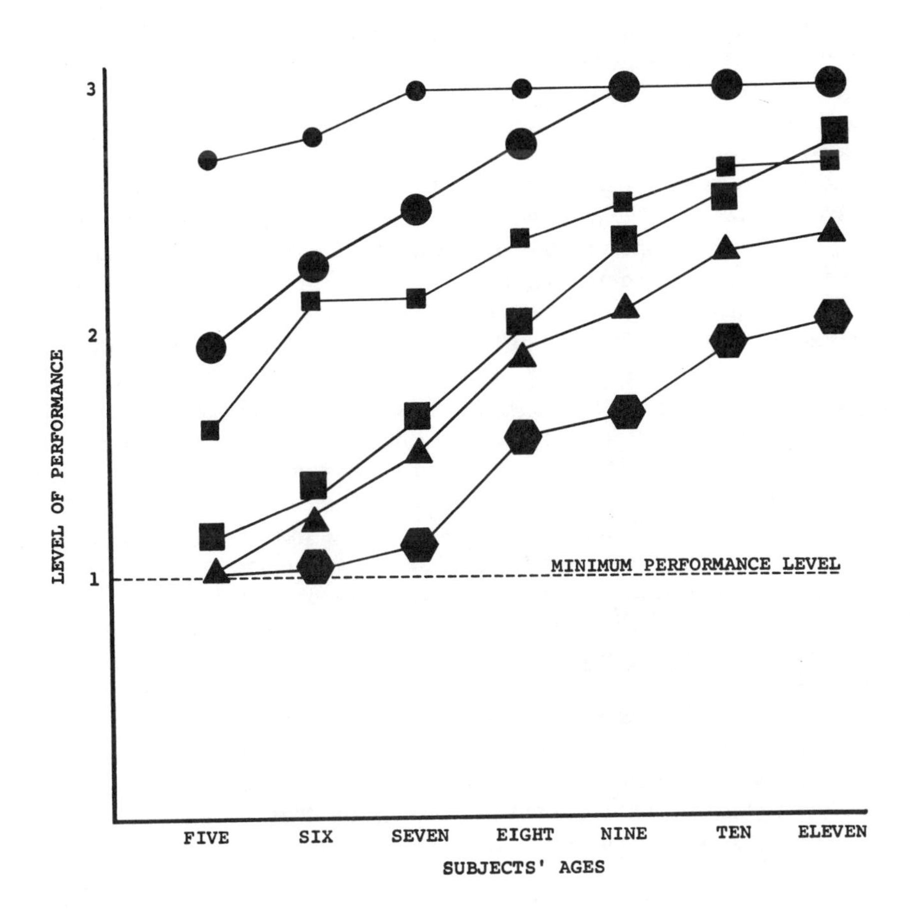

Source: Constructed by the author.

subject grasp the connection between type of correspondence and relative manyness.

Order of Emergence of the Six Concepts

To determine what specific developmental lags obtained between the six concepts, the general procedure discussed in Chapter 8 was employed again. That is, subjects' levels of performance on a given concept was compared with their level of performance on the remaining five concepts. As noted above, each subject had already been assigned one of the three levels for each concept. To compare a given concept (for example, class extension) with some other concept (class inclusion), it remained only to cross-classify all 350 subjects according to their levels of both concepts. With six concepts, there are a total of 15 possible cross-classifications. The results of all 15 cross-classifications appear in Table 10. 2. The binomial formula again was used to determine whether or not a given cross-classification revealed a statistically significant developmental lag. Thus, for example, if we wish to determine whether or not there is a significant lag between class extension and class inclusion, we begin by counting all the subjects who were assigned to higher levels of class extension than class inclusion. (Table 10. 2 indicates 226 in all.) Next, we count all the subjects who were assigned to higher levels of class inclusion than class extension. (Table 10. 2 indicates 13 in all.) Finally, the binomial formula is used to test the ratio 226/239 (the ratio of the number of subjects functioning at higher levels of class extension to the total number of subjects whose class-extension and class-inclusion levels were discrepant) for statistical significance. In this particular case, the binomial formula shows that the probability that class extension does not actually appear in children's reasoning before class inclusion is vanishingly small. The same is true for each of the remaining 14 comparisons. Of course, the specific probability values given by the binomial formula were never the same for any two comparisons. However, all 14 values were vanishingly small. To illustrate, the largest value occurred for the comparison of cross classification and cardination. Even for this comparison, the binomial formula indicated that there was less than one chance in ten thousand that the observed ratio (80/98) could have occurred by chance alone. In short, therefore, the binomial tests of the cross classification data appearing in Table 10. 2 strongly suggested that there are developmental lags between all six of the concepts. Let us now briefly consider what the specific lags are.

For convenience of both examination and exposition, the cross-classification data on the six concepts have been arranged in Table 10. 2 according to the order in which children seem to acquire them.

TABLE 10.2

Developmental Relationships Between Cardinal Concepts

	Class concept			Cardinal equivalence			Class extension			Cross classification			Cardination			Class inclusion		
	I	II	III	I	II	III	I	II	III	I	II	III	I	II	III	I	II	III
Class concept																		
I				4	0	0	4	0	0	4	0	0	4	0	0	4	0	0
II				10	2	0	12	0	0	12	0	0	12	0	0	12	0	0
III				24	40	270	53	106	175	89	139	106	117	152	65	196	100	38
Cardinal equivalence																		
I	4	10	24				34	4	0	38	0	0	38	0	0	38	0	0
II	0	2	40				17	20	5	33	9	0	42	0	0	42	0	0
III	0	0	270				18	82	170	34	130	106	53	152	65	132	100	38
Class extension																		
I	4	12	53	34	17	18				61	8	0	64	5	0	69	0	0
II	0	0	106	4	20	82				29	66	11	37	56	13	86	7	13
III	0	0	175	0	5	172				15	65	95	22	91	52	47	93	25
Cross classification																		
I	4	12	89	38	33	34	61	29	15				91	12	2	98	7	0
II	0	0	139	0	9	130	8	66	65				33	102	4	78	44	11
III	0	0	106	0	0	106	0	11	95				9	38	59	36	49	27
Cardination																		
I	4	12	117	38	42	53	64	37	22	91	33	9				133	0	0
II	0	0	152	0	0	152	5	56	91	12	102	38				57	74	21
III	0	0	65	0	0	65	0	13	52	2	4	59				22	26	17
Class inclusion																		
I	4	12	196	38	42	132	69	86	47	98	78	36	133	57	22			
II	0	0	100	0	0	100	0	7	93	7	44	49	0	74	26			
III	0	0	38	0	0	38	0	13	25	0	11	27	0	21	17			

Source: Compiled by the author.

Taking the concept of class first, children grasp this particular concept before any of the others. In the entire sample of 350 children, there was not a single subject who was assigned a lower level of the class concept than some other concept. By way of contrast, the numbers of subjects who were assigned higher levels of the class concept ranged from 74 in the case of the class concept as opposed to cardinal equivalence to 308 in the case of the class concept as opposed to class inclusion. Hence, there is overwhelming evidence that the class concept precedes each of the other five notions in children's thinking. In view of the previously noted fact that the class concept is the only one which does not involve either manyness or correspondence, this finding seems perfectly reasonable. Table 10. 2 suggests that cardinal equivalence is the second of the six concepts to appear, emerging after the class concept and before class extension, cross-classification by manyness, cardination, and class inclusion. As far as the latter three concepts are concerned, no subject was assigned to a lower level of cardinal equivalence whereas 197, 247, and 274 subjects, respectively, were assigned to higher levels of cardinal equivalence. Class extension is the third concept to appear; children understand both the class concept and cardinal equivalence before they make much progress with class extension, cross-classification by manyness, cardination, or class inclusion.

Cross classification by manyness, cardination, and class inclusion are the last three concepts to emerge. According to Table 10. 2, children understand the class concept, cardinal equivalence, and class extension before cross-classification, and cross-classification is understood before either cardination or class inclusion. Finally, concerning cardination and class inclusion, these are the last two concepts to emerge. It also appears that cardination precedes class inclusion. Although the developmental lag between these two concepts is not as pronounced as some of the others shown in Table 10. 2, there is, nevertheless, credible evidence that cardination is the first to be acquired. A total of 105 children were assigned to higher levels of cardination than class inclusion. Only 21 subjects were assigned to higher levels of class inclusion than cardination. Of these latter 21 subjects, it is interesting to observe that all of them were assigned to level II of cardination and level III of class inclusion. This particular finding would seem to suggest that children must at least attain an intermediate level of cardination at which they base relative manyness judgments on relative density before they can make any progress with the manyness relationship between superordinate classes and subordinate classes.

SYNOPSIS

To conclude this chapter, let us return briefly to the two questions which constitute the chief motivation for this investigation. At what age, generally speaking, may we say that the connection between the type of correspondence which obtains between two collections and their relative manyness is understood by virtually all subjects? Is it true that cardinal-number concepts tend to emerge gradually throughout the entire elementary school age range? The findings reported in Table 10.1, Figure 10.4, and Table 10.2 provide clear answers to both questions. Concerning the age-norm question, it is reasonable to conclude that the correspondence-to-manyness relationship is never understood by more than a minority of elementary school children. The series of tests administered to the present subject sample included two specific tests, cardination and class inclusion, the solution of which would seem to require that a subject understand that relative manyness can be deduced from type of correspondence. By and large, elementary school children performed very poorly on both of these tests. Even at the two oldest age levels (10 and 11 years), less than a majority had attained level III of cardination and class inclusion. On the basis of test performance, only 34 percent of the ten-year-olds and 48 percent of the eleven-year-olds appeared to be capable of complete internal correspondence. These same subjects' performance on the class-inclusion tests was even worse. Only 30 percent of the ten-year-olds and 34 percent of the eleven-year-olds clearly understood that a superordinate class always contains more terms than any one of its constituent subordinate classes. However, if we consider only the cardination test—which, as we saw in Chapter 8, is probably a more sensitive measure of the correspondence-to-manyness relationship—then our most reasonable conclusion is that, at most, roughly one child in two will understand this relationship by 11 years of age. It seems that we also must conclude that the age level at which virtually all subjects may be expected to grasp the correspondence-to-manyness relationship is somewhere beyond the elementary school years. It is not possible to be more precise than this with the present data. Fortunately, the group of investigators at the University of Wisconsin who conducted the replication study reported in Chapter 8 have also gathered some data which allow us to be more precise about when most children are capable of complete internal correspondence. F. H. Hooper and his co-investigators administered class-inclusion problems very similar to those described earlier to large numbers of junior high, high school, and college students. They found that it was not until the second half of the high school years that the subjects invariably understood that a superordinate class contains more terms than any of its subordinate classes. In view of the fact that, as we saw at the outset of this chapter, the class-inclusion principle appears to be a generalization of the concept of cardination, it seems reasonable to suggest that the age at which virtually all subjects may be expected to be capable of complete internal correspondence probably lies somewhere in the middle of adolescence. As we shall see in the next chapter, this suggestion is extremely surprising when viewed from the perspective of entrenched beliefs about children's cardinal-number skills which currently are prevalent in North American educational circles. Moreover, this suggestion carries with it the implication that some radical changes in our current approaches to elementary school arithmetic instruction are called for. The next chapter will consider what some of the more obvious changes are.

Concerning the second question, the data we have considered are consistent with the ordinal theory's hypothesis of a protracted period of emergence for cardinal-number skills, and they are inconsistent with the contrasting hypotheses of the cardinal and cardinal-ordinal viewpoints. As far as the cardinal theory is concerned, we do not anticipate much in the way of age changes in cardinal skills during the elementary years and, consequently, we do not expect much in the way of developmental lags among such concepts. After all, most of the developmental changes in cardinal concepts are supposed to occur before children enter elementary school. It goes without saying that, empirically, this position is untenable. Massive age changes and unequivocal developmental sequences were observed for the present cardinal concepts. Concerning the cardinal-ordinal theory, we would expect that the great preponderance of age variation and developmental lags would be observed in subjects below the age of eight. With nine-, ten-, and eleven-year-olds, who are well beyond the age for the acquisition of Piaget's concrete-operational thinking structures, neither substantial age changes nor clear developmental lags are anticipated. At the very least, the cardinal-ordinal theory leads us to expect that there should be a pronounced decrease in the observed age variations and lags between the first and second halves of the elementary school years. There is not the slightest support for these expectations. Just as much improvement in cardinal skills occurred in the second half of the elementary school age range as in the first. Improvements in cross-classification by manyness, cardination, and class inclusion, in particular, were restricted almost entirely to age eight and above. For all intents and purposes, no real progress in understanding the correspondence-to-manyness relationship is made during the first half of the elementary school age range. It would seem, therefore, that the extended period of cardinal-number development—spanning at least the elementary grades and perhaps extending beyond—envisioned in the ordinal theory is, at least currently, the most defensible of the three views outlined at the beginning of this chapter.

11

EPILOGUE AND EDUCATIONAL POSTSCRIPT

We have come a rather considerable distance in a short space. At the outset of this book, we considered what recorded history tells us about the cultural evolution of the number concept. We also examined some speculations about the prehistorical origins of the number concept which the early cultural history of numbers make highly probable. We saw that, historically, Western civilizations have espoused two general attitudes toward numbers. First, numbers have been viewed as supersensible Platonic entities which the human mind, in some altogether unfathomable manner, "discovered." This view, which was almost universally accepted from the dawn of recorded history until the middle of the preceding century, strongly suggests that numbers originated, during prehistorical time, as objects of religious worship. Whatever its other merits may be, this metaphysical conception of numbers leads unfailing to numerology and its associated eccentricities. The hoaxes perpetrated by medieval numerology were so improbable that men of common sense, such as Galileo and Bruno, were ultimately led to reject numerology itself. During the Renaissance and the Age of Reason, the rival hypothesis that numbers, like gun powder, were invented to fulfill specific needs took shape. However, it was not until the final quarter of the nineteenth century that this hypothesis was made more credible than the Pythagorean view by the advent of logical theories of the number concept. With the appearance of these logical theories, Pythagoreanism fell from favor and, with the possible exception of intuitionism, it has remained so to the present day.

There can be no doubt that Pythagoras' hypothesis that numbers are supersensible entities which hold the key to unlocking the secrets of the universe, and the superstitous veneration of numbers associated with this hypothesis, are chiefly responsible for the privileged position accorded the number concept even today. A large part of elementary school education, for example, is concerned with teaching numerical computation. We deem it essential that, to merit the adjective "civilized," a person be able to read and write, on the one hand, and to use numbers, on the other. The latter requirement is inexplicable from either a logical or utilitarian point of view. As far as logic is concerned, number is not an especially important notion. It is neither more nor less important than other mathematical ideas. Logic is capable of defining numbers, as it is capable of defining other mathematical concepts, and logic tells us that the role which numbers fulfill in computation can be fulfilled by other constructs (for example, relations or classes). As far as utility is concerned, it must be admitted that numbers are useful devices for meeting certain demands of civilized life, especially life in modern technological societies. But there are other equally important mathematical ideas—in particular, the measurement operations of geometry—which appear just as useful as numbers. And yet we would never think of according them the same status as number. As much as we may be loath to admit it, the peculiar position occupied by numerical skills in the hierarchy of intellectual accomplishments valued by civilized cultures becomes understandable only when we take account of Pythagorean superstition. This leads to a somewhat gloomy observation: Discredited superstitions, unlike discredited scientific theories, have a way of reappearing from time to time.

In Part I, we saw that modern mathematical logic has produced two rather different answers to the question, "What are numbers?" Historical priority goes to the answer given by Richard Dedekind. Dedekind identified the natural numbers of arithmetic with the ordinal numbers which, in turn, may be identified with simple infinite progressions. Although the specific terms of any two progressions may differ, they must always have one thing in common; that is, their individual terms are given a unique and fixed order by a transitive-asymmetrical relation of some sort. Thus, numbers are treated as nothing more than values to be found in the domains of transitive-asymmetrical relations, and the logic of the number concept becomes synonymous with the logic of transitive-asymmetrical relations. A second tradition, which began with Gottlob Frege and Bertrand Russell, identified the natural numbers of arithmetic with cardinal numbers. In this approach, the transitive-asymmetrical relations of the ordinal tradition were replaced by correspondence relations. Correspondence relations were grouped into two general types—similar and dissimilar—and numbers were defined as classes of all those classes between which the first of the two types of correspondence may be shown to obtain.

Part II examined three psychological theories of number. We saw that the principal goal of any psychological theory of number is to specify the conceptual skills which are essential prerequisites for acquiring arithmetic. There are differences of opinion regarding the psychological roots of arithmetic which, roughly speaking, parallel differences of opinion regarding the logical foundations of arithmetic considered in Part I. However, unlike the logical evidence considered in Part I, the empirical evidence reviewed in chapters 8, 9, and 10 provided strong support for one particular psychological theory, the ordinal theory. Both the developmental and functional predictions of the ordinal theory received consistent confirmation, while the predictions of two contrasting theories were refuted with equal frequency. In the absence of either contradictory data or logically compelling counterarguments, one can therefore provisionally conclude that the human number concept, as indexed by arithmetic competence, initially evolves from a prior understanding of ordinal number and not from a prior understanding of cardinal number. The cumulative support provided for this conclusion by the empirical findings which we have considered is sufficiently unequivocal so as not to require further comment. Therefore, the balance of the present chapter will be devoted to the exploration of an entirely new question which arises from this conclusion. The question concerns some implications of the ordinal theory and the data which support it for elementary school arithmetic instruction.

THE NEW MATHEMATICS

Since we initiated our inquiry with historical remarks, perhaps it is only fitting that we close with some. By comparison with our earlier consideration of the cultural history of numbers, the events which will be taken up here occurred very recently and, in consequence, are more accessible and less subject to doubt. In fact, many of them fall within the personal experience of most readers. The events in question are those surrounding the advent of the so-called "new school mathematics" (or, more colloquially, the "new math") in the elementary school classrooms of North America. A thoroughly engaging account of this period has been written by D. A. Johnson and R. Rahtz (1966). The historical comments about the birth of the new school mathematics which follow are derived primarily from this source, but the criticisms and counterarguments, quite obviously, are not.

During the late 1950s and early 1960s, a new philosophy of mathematics instruction won almost universal acceptance in public school systems of the United States and Canada. Those who remember this era well—either as parents trying to cope with unfamiliar mathematical concepts or as public school students wondering what all the furor was about—no doubt will also recall the rallying cry of the approach—"new math." The new math held out the promise of creating new generations of adults whose arithmetic and algebra skills far surpassed those of current or preceding generations. However, it recently has become all too clear that the new math has failed to deliver on this promise. Objective evidence suggests that there has been a precipitous decline in mathematical competence of public school students since the introduction of the new math, and many of its advocates are searching for an explanation for this disconcerting phenomenon.

The explicit problem to be considered, then, is this: Why is it that the new school mathematics has failed to deliver, and why, more particularly, do children who have been subjected to this approach turn out to be poorer mathematicians than the pupils of a generation ago? To some readers, this may seem a rather strange question to pose at the end of a book which is nominally concerned with more theoretical matters. As it turns out, however, the studies presented in the preceding three chapters, whose original motivation was simply to test three competing views of number development, happen to entail a very reasonable explanation of why the new school mathematics has failed. Even more important, some of the findings of these studies suggest a procedure whereby we might remedy this situation. Unfortunately, the connection between these data and the failure of the new math to achieve its stated objectives is not apparent at first glance. To see precisely what the connection is, we must first examine the new school mathematics itself.

Origins and Content

Until the late 1950s, public school mathematics education focused on mathematical discoveries made prior to 1800. Elementary school and junior high school mathematics were confined to arithmetic and its applications. During the first few elementary grades, children were taught to add and then to subtract the first ten natural numbers. During the later elementary grades, these two operations were extended to larger numbers, and multiplication was introduced as a generalization of addition (that is, $(x) \cdot (y)$ means x added to itself y times) and division was introduced as a generalization of subtraction (x/y means how many times can y be subtracted from x). Fractions were introduced during grade six. One extremely interesting feature of elementary school arithmetic during this era is that there was no distinction between numbers and numerals. That is, children were not explicitly taught either the ordinal or cardinal meanings of the symbols "1," "2," "3," . . . It was assumed that children enter elementary school with a native grasp of what these symbols mean and, therefore, we may proceed immediately with $1 + 1 = 2$, $1 + 2 = 3$, and so on.

In the seventh and eighth grades, much time traditionally was spent on fractions, decimals, and long division. Story problems were stressed. Seventh and eighth graders also were taught everyday and business applications of arithmetic (determining the interest on loans, converting one currency into another, balancing assets and liabilities). During high school, ninth- and eleventh-grade mathematics were devoted to algebra. Tenth-grade mathematics was devoted to plane geometry. Twelfth-grade mathematics was devoted to solid geometry and trigonometry.

A movement to revamp public school mathematics developed among American academic mathematicians and educators during the early 1950s. The nominal rationale for this movement was that the technological explosion which was then taking place (the transistor, for example, had just been invented) made it absolutely essential to broaden the citizenry's level of mathematical facility. Different groups within the movement emphasized different ways of achieving this altogether admirable goal. As it might be expected, academic mathematicians, who know a great deal about mathematics but very little about the practical problems involved in imparting it to children, tended to stress the need for bringing the "content" of public school mathematics up to date. In particular, they advocated the introduction of such new topics as symbolic logic (especially the theory of classes) and non-Euclidean geometry. It was argued that room could be made for these new topics without eliminating any traditional subject matter by simply reducing the amount of time spent on the application of mathematics to everyday problems. As also might be expected, academic educators, who know next to nothing about mathematics but are well-schooled in the problems of teaching children, tended to stress the need for innovations in the "methodology" of mathematics instruction. They advocated two specific innovations. First, it was proposed that mathematics should be taught via the so-called "self-discovery" method. Self-discovery instruction essentially is the ancient Socratic method. The classic illustration of how method works, an illustration which is quite apropos considering that it deals with mathematics, appears in Plato's Meno. Therein, Socrates causes an illiterate boy to "discover" the Pythagorean theorem by posing leading questions. The assumption that answering leading questions is in some meaningful sense self-discovery and the assumption that self-discovery is the best way to learn mathematics are both open to serious doubt. However, this is the method which was and is advocated. Second, mathematics educators sought to effect a considerable reduction in the amount of time spent on mathematical drill. They proposed that drill, which they viewed as unacceptably tedious, should be partially replaced by instruction concerned with the underlying meaning of concepts on which children simply had been drilled in the past. It was argued that a child who receives a modest amount of drill with, for example, division and who also receives instruction in the meaning of the division concept ultimately will be able to divide better than a child who simply receives a very large amount of division drill. Although this argument is a comfortable tenet of faith, it is, like the argument for self-discovery instruction, open to serious doubt. However, since the present critical observations will be confined to questions concerned with the content of public school mathematics, we shall not pursue this argument at any length. We can simply note that, although we may not wish to acknowledge it, it nevertheless is true that we know of no psychological principle which establishes that someone who "understands" a certain concept invariably will be able to apply it with greater facility than someone who has simply learned the specific applications of the concept.

As one consequence of the movement to reorganize mathematics education, a series of experimental curriculum projects were undertaken at various universities in the United States. Among the most influential of these projects were the University of Maryland Mathematics Project, which began in 1957, the School Mathematics Study Group, which began in 1958 at Yale University, and the University of Illinois Committee on School Mathematics, which began in 1952. Different experimental projects focused on different aspects of public school mathematics. For example, some projects focused exclusively on elementary school mathematics, while other projects concentrated on the junior high and high school grades. Different projects concerned with the same grades offered somewhat different instructional and content recommendations. However, there was broad agreement among all experimental projects on two fundamental and closely related goals. First, it was decided that the meaning of mathematical concepts should be stressed and drill should be deemphasized. Second, regarding content changes, it was decided that the logic of classes—or "set theory" as it is more commonly called by teachers—would be adopted and taught as the universal foundation of mathematics. In particular, the logic of classes was to be adopted as the first step in elementary school arithmetic programs. It is this particular feature of the new school mathematics which will concern us.

As every parent knows, it is the heavy emphasis on the logic of classes that serves to distinguish contemporary arithmetic instruction from earlier arithmetic programs. This fact is readily acknowledged by mathematics educators. For example, Johnson and Rahtz observe that "one of the hallmarks of the new school mathematics—we can almost say _the_ hallmark—is the topic of sets" (1966: 35). It is precisely this reliance on the logic of classes which may account in large measure for the failure of the new approach to live up to its advance billing, or so it will be argued later on. First, however, let us consider both why and how the theory of classes is introduced in the new school mathematics.

Concerning the "why" question, the sole justification for incorporating the logic of classes into arithmetic instruction is to achieve the aforementioned objective of imparting to children the meaning of relevant concepts. In other words, the logic of classes was adopted as a source of basic ideas, which virtually all children presumably understand by the time they enter elementary school, that teachers can draw on to define the elementary concepts of arithmetic. Although the logic of classes is used to define all the important ideas of arithmetic in the new school mathematics, it was first proposed on the ground that it provides a mechanism for teaching children the meaning of the number concept (see Johnson and Rahtz 1966: Ch. 1). It was noted earlier that the meaning of the number concept was not explicitly taught to elementary school children before the advent of the new school mathematics. The blatantly Pythagorean assumption was made that children possess an innate grasp of what "1," "2," "3," . . . mean and, therefore, mathematics education can begin with learning how to combine these symbols properly. In the new school mathematics, on the other hand, children are first taught the meaning of these symbols. As might be expected, the specific meaning that is taught is the cardinal theory. Teaching the cardinal meanings of "1," "2," "3," . . . is the very first step in arithmetic instruction. All subsequent topics (addition, subtraction, multiplication, and division) are based on this initial encounter with cardinal number. For the sake of perspicuity, we should examine precisely how arithmetic concepts are founded on the cardinal theory.

Examples were selected from what is currently the most widely used commercial version of the new school mathematics in North America (Eicholz 1969). The program begins, at the kindergarten and first-grade levels, with the idea of class. That is, children are taught that single objects may be grouped into classes on the basis of shared intensional properties (for example, rabbits are a class, earthworms are a class, carrots are a class). After dealing with the concept of class, the relationship between manyness and correspondence is introduced. Children are taught that the relative manyness of two classes can always be determined by ascertaining the type of correspondence between them. For example, children may be shown that, given a collection of squirrels and a collection of walnuts, the way to determine the relative manyness of the two collections is to construct a term-by-term correspondence. If there is one squirrel for each and every walnut, then the classes contain equally many terms; if there are some squirrels remaining after each walnut has been assigned to a squirrel, then the squirrels are more numerous; if there are some walnuts remaining after each squirrel has been assigned to a walnut, then the walnuts are more numerous. The concrete procedure which children are taught to employ to establish such correspondences is that of drawing a straight line from each term of one class to a given term of the other class. Many everyday class pairings of the sort just described (dogs and dog houses, frogs and lily pads, shoes and socks) are used to illustrate and inculcate the idea that relative manyness depends on type of correspondence. As can easily be seen, the aim of this sort of instruction is to induce the concepts of similarity and dissimilarity of classes.

Obviously, the next step is to teach children the cardinal theory itself. The theory is taught in two phases. First, the notion of "absolute manyness" is presented. This is done by teaching children that, in addition to being able to group collections together on the basis of shared intensions (for example, automobiles and trucks may be grouped together because both are modes of transportation), collections may be grouped together on the basis of their extensions. Exercises designed to inculcate this fact include selecting pairs of classes containing equally many terms from among several alternative classes. For example, the child might be shown a target class containing a pair of baseball bats and three alternative classes containing a single glove, a pair of balls, and a trio of caps. The task is to determine which of the latter classes contains exactly as many terms as the target class. The second step in teaching the cardinal theory is similar in principle to condition 3 of the second experiment reported in Chapter 9. Numerals are introduced as symbols denoting particular absolute manynesses. For example, the numeral "2" is defined by showing the child classes containing a pair of blocks, a pair of shoes, and a pair of books. He or she is then expected to learn that the symbol "2" always denotes that particular manyness which adults call "a pair." Further exercises designed to promote the same idea involve, much as in the third condition of the aforementioned experiment, learning to chose which class of several possible classes should be denoted by the numeral "2." For example, the child might be shown classes containing a pair of Indians, a single canoe, a pair of teepees, and a trio of drums, respectively, and the task would be to select the specific classes for which the symbol "2" is appropriate. Each of the first few natural numbers is introduced in this same general way, and they are introduced in order. Thus, the child first is taught that "1" goes with classes containing a single term; next, the child is taught that "2" goes with classes containing a pair of terms, and so on. By the end of the first grade, children are expected to understand the cardinal meanings of the first ten natural numbers. Consistent with the cardinal theory of number, the overriding aim is to teach the child that each of these numbers is a particular classes of classes: that is, that class whose members are classes containing a specific absolute manyness of terms.

Instruction of this type is not completed until roughly the middle of grade one. When it is completed, addition is presented for the first time. Naturally, addition is introduced as a procedure whereby classes of greater manyness are created by uniting classes of lesser manyness. Hence, children are taught that "1 + 1 = 2" means that a class containing a single term united with another class containing a single term yields a class containing a pair of terms, similarly, "1 + 2 = 3" is said to mean that a class containing a single term united with a class containing a pair of terms yields a class containing a trio of terms, and so on for manynesses up to a decade. Note that this particular definition of addition is essentially the same as the principle of class inclusion examined in the preceding chapter. Following addition, subtraction is introduced as the reverse of addition—that is, as a method whereby classes of lesser manyness can be derived from classes of greater manyness. On the whole, therefore, the aim of addition and subtraction instruction is to found both notions squarely on the cardinal meaning of numbers.

By the end of grade one, children are expected to at least be able to add the first five natural numbers correctly. It is also hoped that they will be able to subtract them correctly. During grade two, addition and subtraction are extended to larger numbers. By the end of grade two, children are expected to be capable of adding and subtracting at least the numbers one through ten. During subsequent elementary grades, multiplication and division are introduced as further extensions of the cardinal theory. Multiplication is characterized as a method for producing a class of greater manyness (the product) from classes of lesser manyness by adding a class of a certain manyness to itself so many times. Hence, "5 x 2 = 10" is taken to mean that a pair of quintuples added to each other yield a decade. Division is characterized as a method for producing a class of lesser manyness (the quotient) from classes of greater manyness by determining how many times a given class can be subtracted from another class. Hence, "6 ÷ 3 = 2" is taken to mean that a trio can be subtracted from a sextuple a pair of times.

To summarize, the original objective of the new school mathematics was to produce palpable improvements in mathematical skill. The specific computational skills that this new approach sought to improve are the same ones stressed in more traditional curricula—addition, subtraction, multiplication, division, fractions, decimals. However, it was hoped that computational accuracy could be greatly increased by altering the content and the methodology of instruction. Explicitly, it was hoped that improvements would result from teaching children the underlying meaning of concepts and operations rather than simply teaching them how to use concepts and operations. The most dramatic changes have been effected in elementary school mathematics.

Traditional arithmetic has been replaced by an arithmetic founded completely on the cardinal meanings of numbers and the cardinal meanings of the four operations of arithmetic. It is assumed that knowing the cardinal meanings of numbers and arithmetic operations will lead to greater proficiency with the operations.

Before examining criticisms of the new school mathematics suggested by the data presented in the preceding three chapters, we consider the curious circumstances which led to its almost universal adoption in North America. In view of the major changes which it recommends, it is somewhat surprising that the new school mathematics won any following at all. It must be admitted, after all, that professional educators are generally a conservative lot, not overly inclined toward curriculum experimentation. Public school officials in particular, owing no doubt to their constant contact with parents, are notoriously intransigent. Thus, at first glance, one would not rate the chances of acceptance of a highly experimental mathematics program as very promising. Nevertheless, the new school mathematics carried the day in less than a decade. Why? As it turns out, fate intervened in October of 1957.

As many remember, the United States and the Soviet Union were engaged in a race to launch the first orbital satellite during the middle 1950s. This also was the period of the so-called "cold war," when every technological achievement of one side was viewed as a blow to the collective ego of the other side. During the summer of 1955, the United States instituted Project Vanguard. The nominal (and well-publicized) aim of the project was to orbit a small satellite (two to three pounds) at a height of roughly 300 miles sometime during 1958. In October of 1957, however, the Soviet Union announced the launching of Sputnik—which means, quite literally, "fellow traveler." Not only had the Soviet Union won the satellite race, certain technological achievements involved in the launch were far in advance of anything anticipated by Project Vanguard. For example, the Soviet satellite was considerably heavier (184.3 pounds), and its orbit was considerably higher (559 miles) than that of its proposed U.S. counterpart.

To say that the American ego, always a sensitive commodity, had been bruised would be something of an understatement. The Soviet satellite program had accomplished things beyond even the most optimistic assessments of its U.S. counterpart, and the American public demanded to know why. In point of fact, the answer was elementary. The Soviet satellite program was allowed to make use of what were, at that time, the most powerful and sophisticated of military rockets. The rocket that eventually launched Sputnik developed over 200,000 pounds of thrust. Project Vanguard, in contrast, was confined to the use of small experimental rockets capable of between only 20,000 or 30,000 pounds of thrust. Unfortunately, the public would have none of simple explanations. The image of American technological omnipotence had been severely tarnished and blame had to be fixed. Ultimately, the largest single share of the blame was placed on the doorstep of public education. Excellence in technology, after all, presupposes excellence in science and mathematics. The reverse of this argument, which, of course, does not necessarily follow, is that technological mediocrity is the result of inadequate science and mathematics education. This latter argument, which was widely discussed at the time, precipitated an unparalleled desire for innovation and experimentation in public school science and mathematics curricula among both professional educators and, more importantly, parents. As far as mathematics curricula are concerned, the new school mathematics was the instrument by which this desire was fulfilled.

Problems and Criticisms

It has become increasingly clear in recent years that the new school mathematics, contrary to its avowed aim, has failed to produce noticeable improvements in the mathematical skills of public school students. Alarmingly, it appears to have had precisely the opposite effect. Since 1969 and 1970, when, for the first time, large numbers of students trained exclusively in the new school mathematics began graduating from high school, there has been mounting evidence of an across-the-board deterioration in mathematical skills. The decline was first noted when these students began entering college. There were two principal sources of evidence. First, most colleges require that applicants, in addition to possessing a high school diploma, take an entrance examination. Among other things, such examinations include a section designed to measure the minimum mathematical skills deemed essential for studies. Students who fail to show these minimum skills subsequently are required to enroll in remedial mathematics courses. Since the late 1960s, as the proportion of college applicants trained exclusively in the new school mathematics has risen, so has the proportion of applicants required to enroll in remedial mathematics courses. The second college index of the decline is nonmathematics courses whose content is especially sensitive to mathematical skill. Introductory courses in chemistry and physics, which have very large enrollments in most colleges, provide perhaps the best illustrations. These courses, usually taken during the freshman year, presuppose that the student has at least a working understanding of such standard algebraic notions as exponents, logarithms, permutations, and systems of equations. They also require modest facility with algebraic manipulation of equations. Since the late 1960s, instructors of such courses have been reporting that an increasing proportion of their students apparently do not possess these prerequisite mathematical skills. The problem has grown so severe of late that some teachers of introductory physics and chemistry courses routinely begin with a refresher unit on basic algebra.

The erosion of mathematical skills has not only been observed in college students. It has now become apparent that high school students also have regressed. As readers who received their elementary and secondary education in North America are aware, standardized achievement tests, which invariably contain mathematics subtests, have been an integral part of high school evaluation for several decades. Some of these tests, usually administered during the eleventh grade, are part of national, state, or provincial scholarship competitions. Many readers, no doubt, will recall having taken the two best-known national tests: the Public School Achievement Test and the National Merit Scholarship Qualifying Test (United States). Overall performance on such tests is always broken down and reported by content area. During recent years, it has been noted that the average performance on the mathematics sections of these tests has been declining steadily. Compared with students taking the same tests a decade and one-half ago, contemporary high school students appear to know far less about algebra, geometry, and trigonometry. This, of course, is quite consistent with the aforementioned findings on college students.

But what about elementary school children? Mathematics is a logically and hierarchically organized body of knowledge. Each subsequent topic in mathematics education (for example, algebra and geometry) builds on some earlier topic (arithmetic). If some topic is poorly understood, then the most probable explanation is deficiencies in some earlier prerequisite topics. Hence, if the algebraic, geometric, and trigonometric capacities of high school and college students are in a state of eclipse, the first explanation which suggests itself is deficiencies in basic arithmetic. Consistent with this line of reasoning, pronounced regressions in the arithmetic skills of North American elementary schoolers have been verified. Elementary school children, like high school students, are administered standardized tests of achievement. The most familiar and redoubtable of these is the Stanford Achievement Test. As is the case with high school achievement tests, standardized elementary achievement tests always include mathematics subtests which, of course, deal with arithmetic skills. Recent results from these tests indicate that the capacity of North American elementary students to add, subtract, multiply, divide, manipulate fractions, and so on, is clearly inferior to the capacity of their counterparts of the previous generation and one-half. The findings prompted the California state department of education to undertake an independent study of elementary school arithmetic skills. They compared the standardized mathematics achievement tests scores of children then enrolled in sixth grade with the scores of sixth graders from previous years. The same disconcerting deterioration of arithmetic skills was observed. Subsequently, this same general finding has been reported by public school authorities in other states and in the province of Ontario in Canada (Laidler 1975).

EXPLANATIONS AND SOLUTIONS

Given the previously discussed goals of the new school mathematics, the degeneration of mathematical skills has been viewed with considerable alarm and has proved a source of severe discomfort to professional educators. Within the past year or so, two very different explanations of these losses have been heard with increasing frequency. On the one hand, an explanation proposed by educational conservatives, who oppose experimental curricula on principle and long for a return to early-1950s curricula, is essentially, "I told you so." More explicitly, it is proposed that both the "new methods" and "new content" of public school mathematics are mistakes. Concerning methods, this explanation has it that teaching students, especially elementary school children, the underlying meaning of a concept is not a better way to learn mathematics. Instead, this strategy tends to confuse students, and it causes them to drift from the straight and narrow path of learning how to add, subtract, multiply, and so on. Hence, we should abandon our illusions about promoting understanding of mathematical concepts and revert to intensive computational drill. Concerning content, the conservative explanation has it that new topics such as the logic of classes, non-Euclidean geometry, and probability should either be deleted entirely or confined to college preparatory curricula. The average citizen, so the argument goes, only needs to know arithmetic and algebra. Since esoteric topics such as the logic of classes and non-Euclidean geometry will be useful to students only if they enter college and then only if they major in science, engineering, and mathematics, there is no justification for borrowing time from arithmetic and algebra instruction to teach these topics.

A second explanation, which is more or less the reverse of the preceding one, comes from advocates of the new school mathematics. According to this explanation, there is nothing wrong with either the methodology or the content of the approach. Instead, the losses in mathematical skill which have been observed of late are symptomatic of shifts in cultural attitudes toward mathematics. During the 1950s and early 1960s, thanks to Sputnik and the cold war, science and technology achieved a previously unparalleled position in the social milieu of North America. The governments of the United States and Canada, by virtue of selective grants to secondary education and colleges, strongly encouraged careers in the basic sciences and engineering. Among the public at large, attitudes toward science and technology were at a high ebb. During the second half of the 1960s, however, the cultural valuation of science and technology moderated perceptibly. There were many reasons for this shift. Two frequently mentioned ones are historical inertia and the environmentalist movement. Concerning the former, after the unrestrained optimism of the 1950s and early 1960s, it was inevitable that some disillusionment with the ability of science and technology to improve the quality of life would set in. By virtue of the environmentalist movement, the public became aware, as never before, that certain technological advances are not automatically beneficial. It has now become axiomatic that there is usually a price to pay in the form of environmental deterioration for major technological achievements. Whatever the reasons, the important fact to bear in mind, so the argument goes, is that cultural valuation of science and technology is rather low at present. Therefore, students simply are not as strongly motivated to learn mathematics as they were a decade or so ago, because they are not contemplating careers for which mathematical knowledge is crucial. This, it must be admitted, is a very face-saving explanation. It leads to the conclusion that we should not make any changes in the new school mathematics because, if there comes a time, as there surely must, when cultural valuation of science and technology is again high, then the new school mathematics will be able to fulfill its promise of creating a generation of mathematically sophisticated citizens.

Neither of the preceding explanations is acceptable. The first explanation is unacceptably reactionary. It is psychologically unsound insofar as its methodological recommendations are concerned. As mentioned earlier, there is no immutable psychological law which states that learning to understand the meaning of a concept must invariably lead to better usage than simply being drilled in usage per se. However, generally speaking, the combination of learning underlying meaning and moderate drill leads to better usage than drill alone. There are specific psychological theories, such as learning set (Harlow 1949) and rule learning (Scandura 1974) that deal with this phenomenon. Moreover, the difference between meaning-plus-drill learning and drill-only learning tends to be especially pronounced for logical and mathematical concepts. Hence, there is good reason to suppose a priori that considerable emphasis should be placed on understanding concepts in public school mathematics curricula. As we shall see below, however, we must also take into account children's levels of cognitive development when we decide which specific meanings to teach and when to teach them.

The second of the preceding explanations is simply a rationalization for ignoring an alarming problem. Moreover, the explanation itself is logically untenable. Although low cultural valuation of mathematics-related careers may have some influence on the motivation of college students and older high school students, it seems extremely unlikely that it has a comparable effect on the desire of elementary and junior high school students to attain excellence in mathematics. Blissful indifference probably is the most reasonable characterization of these students' attitudes toward the relevance of mathematics to their ultimate niche in the adult world of work.

A third hypothesis may be proposed about the failure of the new school mathematics. This hypothesis is simpler than the two which we have already considered, and it is given prima facie credibility by the data on children's numerical concepts reported in chapters 8, 9, and 10. To begin with, it assumes, in line with earlier remarks, that the decline in algebraic, geometry, and trigonometric skills at the high school and college levels is to some extent a consequence of the corresponding decline in arithmetic skills among elementary schoolers. If children cannot multiply and divide natural numbers with facility, then their chances of acquiring more advanced skills which presuppose such facility (fractions, decimals, exponents, logarithms) are not at all promising. To explain the failure of the new school mathematics, therefore, we must explain the decline in elementary school arithmetic skills.

Although there are undoubtedly other contributing factors, the emphasis placed on the logic of classes probably is at least partially responsible for the decline in arithmetic skills. It will be recalled that ideas from the logic of classes, especially those notions involved in the Frege-Russell theory of number, are used to define all the basic concepts of arithmetic. To understand the explicit definitions of arithmetic concepts such as number, addition, and multiplication, it is self-evident that the child first must understand the logical ideas on which these definitions are predicated. As we saw with the illustrative curriculum considered earlier on, first-grade children must at least grasp the ideas of class, class extension, relative manyness, term-by-term correspondence, the manyness-to-correspondence relationship (similarity and dissimilarity), absolute manyness, and class inclusion. If they do not understand these ideas, then the definitions which make use of them obviously will be incomprehensible and the curriculum will be futile. Moreover, given the inevitable frustration and disaffection generated by instruction that is beyond one's present level of cognitive sophistication, the long-range impact of such a curriculum on arithmetic skills presumably would be negative. Other things being equal, therefore, it would be far better to leave arithmetic concepts undefined than to propose definitions which cannot be understood by the average child.

If the findings reported in preceding chapters show anything at all, they show that most children clearly do not comprehend the logical ideas used to define arithmetic concepts in the new school mathematics. In fact, of the specific logical ideas just mentioned as playing important roles in grade-one instruction, the concept of class is the only one which, it seems safe to conclude, the great preponderance of five- and six-year-olds understand. None of the remaining ideas are understood until substantially later. This is especially true of the manyness-to-correspondence relationship and the class-inclusion principle. We saw in the earlier curriculum illustration that both of these ideas play a central role in definitions of number and the four arithmetic operations performed on numbers. The definitions of number, addition, and subtraction which make use of these notions are introduced during the first half of grade one. Concerning number in particular, the definition of the generic number concept involves the manyness-to-correspondence relationship and the definitions of individual numbers involve the class inclusion principle. By the end of grade one, children are expected to know both the generic number concept and the specific cardinal meanings of the first ten numbers.

We saw in Chapter 10 that the available normative data certainly do not lead one to suppose that the manyness-to-correspondence relationship and the class-inclusion principle are understood by the great majority of first graders. (For that matter, the data suggest, neither are the somewhat easier concepts of class extension, relative manyness, correspondence, and absolute manyness.) Quite to the contrary, first graders do not show the slightest evidence of either notion. Further, our best evidence is that, even by grade six, the manyness-to-correspondence relationship and the class-inclusion principle are not comprehended by more than a minority of children. It appears that adolescence is the earliest age level for which we may state with some confidence that virtually all students will grasp both ideas. And yet, by the middle of grade one, the new school mathematics expects that children will have learned a series of definitions which appeal to these ideas. In view of the normative data, it seems highly improbable that first graders actually are learning these definitions. This conclusion is strengthened by the fact that all of the children who participated in the normative study reported in Chapter 10 were pupils of elementary schools in which new school mathematics curricula were in use. Hence, all of the children had received considerable instruction in the concepts measured. Given that the definitions are not understood, a decline in the arithmetic skills of elementary students is virtually guaranteed.

The general supposition, therefore, is that a specific "content variable"—not the overriding emphasis on understanding the meaning of concepts—probably is responsible for the deterioration observed in elementary schoolers' arithmetic during recent years. Further, given what we now know about the difficulty of cardinal ideas for children in the elementary grades, it would be more surprising if the new school mathematics had failed to produce such deterioration. This brings us to the question of how best to restructure public school mathematics curricula to increase skills to previous levels and, perhaps, beyond. There are two obvious possibilities. First, we can reject both the methodological and content innovations of the new school mathematics and revert to the drill-oriented instruction that was popular a decade and one-half ago. There seems to be no scholarly merit in this solution, but it has great practical advantages. The most noteworthy advantage of reverting to the public school mathematics of the 1950s is that it would require very little effort on our part. No time need be spent in the development of new curricula. Programs of this sort already exist, and we have only to put them back into operation. The second solution, which is favored here, involves retaining the methodology of the new school mathematics and altering its content. This solution involves, on the one hand, retaining the emphasis on understanding concepts combined with moderate amounts of drill and, on the other hand, replacing the logical ideas currently used to define arithmetic concepts with logical ideas that are more attuned to children's cognitive development.

By now, most readers will have guessed the specific content changes that would be recommended here. Instead of founding arithmetic and the rest of mathematics on the inordinately difficult ideas of the logic of class, it is urged that we create a "newer new school mathematics" in which arithmetic is founded on the logic of relations. In particular, we should teach elementary school children (1) the ordinal definition of the generic concept of number, (2) the specific ordinal meanings of the first several natural numbers, and (3) the specific ordinal meanings of the four operations of arithmetic. The ordinal definition of the number concept was discussed in sufficient detail in Chapter 3 so as not to require further discussion here. The defining of individual numbers ordinally for children would simply consist of teaching them that each number refers to a unique position in all ordered progressions (rather than to a certain absolute manyness). Similarly, defining specific arithmetic operations ordinally for children consists of teaching them that each operation is a procedure whereby new positions in ordered progressions may be obtained from previously given positions: Addition becomes synonymous with obtaining a later term by combining earlier terms; subtraction becomes synonymous with obtaining an earlier term by combining later terms; multiplication becomes synonymous with obtaining a later term by repeated addition of an earlier term to itself; division becomes synonymous with obtaining an earlier term by subtracting a certain term from a later term a given number of times. In short, the initial aim would be to construct a logic-of-relations counterpart of the illustrative arithmetic curriculum considered earlier.

The merit of the second solution is twofold. First, it allows us to retain the emphasis on understanding which is the chief methodological innovation of the new school mathematics. Second, the solution involves the adoption of logical notions which are known to be part of children's mental repertoires at the time they enter elementary school. The normative data reviewed in Chapter 8 indicate that the logical ideas involved in the ordinal definition of number would be quite comprehensible to five- and six-year-olds. We saw that even kindergarteners, for the most part, possess completely internal concepts of ordering and ordinal number. The experimental data considered in Chapter 9 further reinforce the recommendation that ordinal ideas replace cardinal ideas as devices for defining arithmetic concepts. In the first experiment reported in Chapter 9, we saw that teaching children basic ordinal concepts has a more pronounced facilitative effect on the growth of arithmetic skills than teaching children basic cardinal ideas. In the second experiment reported in Chapter 9, we saw that learning to associate a numeral with its concrete meaning is markedly easier when the meaning in question is ordinal rather than cardinal. Taken together, the findings of these two experiments suggest that, in addition to returning arithmetic skills to their earlier levels, a newer new school mathematics based on the logic of relations and the ordinal theory of number may, at last, permit us to achieve the goal of general improvement in mathematical skills at all age levels.

Even if the data on deterioration of arithmetic skills and the data on the difficulty of cardinal ideas did not exist, it still would be preferable, for purely logical reasons, to found arithmetic on the logic of relations as opposed to founding it on the logic of classes. It will be recalled from Part I that the cardinal definition of number is of absolutely no importance in the rest of classical mathematics. Thus, when we found arithmetic on cardinal ideas, we are not teaching children anything that they can "transfer" to, for example, algebra or geometry. The situation is reversed for the ordinal definition of number. As we saw in Chapter 3, all of classical mathematics springs from the notion of progression—or, more precisely, the notion of transitive-asymmetrical relation. If arithmetic is founded on such concepts, we not only are teaching children definitions they can comprehend; we also are teaching them definitions which will be very useful in all areas of public school mathematics and beyond.

To summarize, the implications of the research examined in Part II of this book extend far beyond the three theories of number development that the research originally was designed to test. In addition to providing consistent confirmation of the predictions of one of the theories and equally consistent disconfirmation of the predictions of the other two theories, the findings suggest a quite plausible explanation of the disconcerting regression in mathematical skills which has been the legacy of the new school mathematics. More important than the explanation, however, is the fact that the findings suggest a positive program whereby educators may perhaps be able to extricate public school mathematics curricula from what has become an increasingly difficult situation. It remains to be seen whether this program will be taken seriously or whether intransigence will continue to prevail.

BIBLIOGRAPHY

Beard, R. M. 1963. "The Order of Concept Development: Studies in Two Fields. I. Number Concepts in the Infant School. Educational Review 15: 105-117.

Beilin, H. and I. S. Gillman. 1967 "Number Language and Number Reversal Learning." Journal of Experimental Child Psychology 5: 263-277.

Bell, E. T. 1940. The Development of Mathematics. New York: McGraw-Hill.

______. 1946. The Magic of Numbers. New York: McGraw-Hill.

Beth, E. W. and J. Piaget. 1966. Mathematical Epistemology and Psychology. Dordrecht, The Netherlands: Reidel.

Benacerraf, P. and H. Putnam. 1964. Philosophy of Mathematics. Englewood Cliffs, New Jersey: Prentice-Hall.

Bingham-Newman, A. M. and F. H. Hooper. 1975. "The Search for the Woozle circa 1975: Commentary on Brainerd's Observations." American Educational Research Journal, 12: 379-387.

Boole, G. 1854. An Investigation of the Laws of Thought. London: Macmillan.

Brainerd, C. J. 1973a. "Mathematical and Behavioral Foundations of Number." Journal of General Psychology 88: 221-281.

______. 1973b. "The Origins of Number Concepts." Scientific American 228(3): 101-109.

______. 1973c. "The Evolution of Number." Paper presented at the Fourth Interdisciplinary Conference on Structural Learning.

______. 1973d. "Order of Acquisition of Transitivity, Conservation, and Class Inclusion of Length and Weight." Developmental Psychology 8: 105-116.

______. 1973e. "Judgments and Explanations as Criteria for the Presence of Cognitive Structures." Psychological Bulletin 79: 172-179.

______. 1975a. "Structures-of-the-Whole and Elementary Education." American Educational Research Journal 12: 369-378.

______. 1975b. "Rejoinder to Bingham-Newman and Hooper." American Educational Research Journal 12: 389-394.

______. 1976. "Response Criteria in Concept Development Research." Child Development 47: 360-366.

______ and F. H. Hooper. 1975. "A Methodological Analysis of Developmental Studies of Identity Conservation and Equivalence Conservation." Psychological Bulletin 82: 725-737.

______ and M. A. Fraser. 1975. "Further Test of the Ordinal Theory of Number Development." Journal of Genetic Psychology 127: 21-33.

______ and P. Kaszor. 1974. "An Analysis of Two Proposed Sources of Children's Class Inclusion Errors." Developmental Psychology 10: 633-643.

Bryant, P. 1974. Perception and Understanding in Young Children. London: Methuen.

Cassirer, E. 1910. Substanzbegriff und funktionsbegriff. Berlin: Wissenschaftliche Buchgesellschaft.

Cohen, M. R. and E. Nagel. 1934. An Introduction to Logic and Scientific Method. New York: Harcourt, Brace, and World.

Dedekind, R. 1887. Was sind und was sollen die zahlen? Braunschweig: Vieweg.

De Long, H. 1970. A Profile of Mathematical Logic. Reading, Mass.: Addison-Wesley.

Dodwell, P. C. 1960. "Children's Understanding of Number and Related Concepts." Canadian Journal of Psychology 14: 191-205.

______. 1961. "Children's Understanding of Number Concepts: Characteristics of an Individual and of a Group Test." Canadian Journal of Psychology 15: 29-36.

______. 1962. "Relations Between the Understanding of the Logic of Classes and of Cardinal Number in Children." Canadian Journal of Psychology 16: 152-160.

Eicholz, R. E. 1969. Elementary School Mathematics, Second ed. Reading, Mass.: Addison-Wesley.

Flavell, J. H. 1970. "Concept Development." In P. H. Mussen (ed.), Carmichael's Manual of Child Psychology. New York: John Wiley and Sons.

Frege, G. 1884. Die grundlagen der arithmetik. Breslau: Marcus.

Gelman, R. 1972. "The Nature and Development of Early Number Concepts." In H. W. Reese & L. P. Lipsett (eds.), Advances in Child Development and Behavior, Vol. 7. New York: Academic Press.

Gödel, K. 1930. "Die vollständigkeit der axiome des logischen funktionenkalküls." Monatshefte für Mathematik und Physik 37: 349-360.

______. 1931. "Über formal unentscheidbare sätze der Principia mathematica und verwandter systeme I." Monatshefte für Mathematik und Physik 38, 173-198.

Gonchar, A. J. 1975. "A Study in the Nature and Development of the Natural Number Concept: Initial and Supplementary Analyses." Technical Report No. 340, Research and Development Center for Cognitive Learning, University of Wisconsin.

Hood, H. B. 1962. "An Experimental Study of Piaget's Theory of the Development of Number in Children." British Journal of Psychology 53: 273-286.

Inhelder, B. and J. Piaget. 1964. The Early Growth of Logic in the Child. London: Routledge & Kegan Paul.

Inhelder, B. and H. Sinclair. 1969. "Learning Cognitive Structures." In P. H. Mussen, J. Langer, and M. Covingtion (eds.), Trends and Issues in Developmental Psychology. New York: Holt, Rinehart and Winston.

Johnson, D. A. and R. Rahtz. 1966. The New Mathematics in Our Schools. New York: Macmillan.

Harlow, H. F. 1949. "The Formation of Learning Sets." Psychological Review 56: 51-65.

Kendler, H. H. and T. S. Kendler. 1962. "Vertical and Horizontal Processes in Problem Solving." Psychological Review 69: 1-16.

Kofsky, E. 1966. "A Scalogram Study of Classificatory Development." Child Development 37, 191-204.

Laidler, K. J. 1975. "The dire effect of the new mathematics on chemical education." Paper presented at the Chemical Institute of Canada, Toronto, Ont.

Lawson, G., J. Baron, and L. Siegel. 1974. "The Role of Number and Length Cues in Children's Quantitative Judgments." Child Development 45: 731-736.

Nagel, E. and J. R. Newman. 1968. Gödel's Proof. New York: New York University Press.

Nelson, T. M. and S. H. Bartley. 1961. "Numerosity, number, arithmetization, and psychology." Philosophy of Science 28: 178-203.

Peano, G. 1894-1908. Formulaire de mathématiques. Vols. 1-5. Turin: Broca.

Piaget, J. 1942. Classes, relations, et nombre. Paris: Vrin.

______. 1949. Traité de logique. Paris: Colin.

______. 1950. The Psychology of Intelligence. New York: International Universities Press.

______. 1952. The Child's Conception of Number. New York: Humanities.

______. 1970a. Genetic Epistemology. New York: Columbia University Press.

______. 1970b. "Piaget's theory." In P. H. Mussen (ed.), Carmichael's Manual of Child Psychology. New York: Wiley.

______ and B. Inhelder. 1956. The Child's Conception of Space. London: Routledge & Kegan Paul.

______ and B. Inhelder. 1969. The Psychology of the Child. New York: Basic Books.

______ and A. Szeminska. 1939. "Quelques expériences sur la conservation des quantités continues chez l'enfant." Journal de Psychologie 36: 36-64.

______ and A. Szeminska. 1941. La genèse du nombre chez l'enfant. Neuchatel and Paris: Delachaux et Niestlé.

Pufall, P. B. and R. E. Shaw. 1972. "Precocious Thoughts on Number Development: The Long and the Short of It." Developmental Psychology 7: 62-69.

Quine, W. V. O. 1951. Mathematical Logic. Cambridge, Mass.: Harvard University Press.

Russell, B. 1903. The Principles of Mathematics. Cambridge, England: Cambridge University Press.

______. 1919. Introduction to Mathematical Philosophy. London: Allen and Unwin.

Scandura, J. M. 1974. Structural Learning. London: Gordon and Breach.

Siegel, L. S. 1971a. "The Sequence of Development of Certain Number Concepts in Preschool Children." Developmental Psychology 5: 357-361.

______. 1971b. "The Development of the Understanding of Certain Number Concepts." Developmental Psychology 5: 362-363.

______. 1974. "Development of Number Concepts: Ordering and Correspondence Operations and the Role of Length Cues." Developmental Psychology 10: 907-912.

Sinclair, H. 1973. "Recent Piagetian Research in Learning Studies." In J. Raph and M. Schwebel (eds.), Piaget in the Classroom. New York: Basic Books.

Smither, S. J., S. S. Smiley, and R. Rees. 1974. "The Use of Perceptual Cues for Number Judgment by Young Children." Child Development 45: 693-699.

Staats, A. W. 1968. Learning, Language, and Cognition. New York: Holt, Rinehart and Winston.

Stabler, E. R. 1953. An Introduction to Mathematical Thought. Reading, Mass.: Addison-Wesley.

Tarski, A. 1939. "On Undecidable Statements in Enlarged Systems of Logic and the Concept of Truth." Journal of Symbolic Logic 4: 105-112.

Taves, E. H. 1941. "Two Mechanisms For the Perception of Visual Numerousness." Archives of Psychology 37: No. 265.

Underwood, B. J. 1966. Experimental Psychology, Second ed. New York: Appleton-Century-Crofts.

Wang, M. C., L. B. Resnick, and R. F. Boozer. 1971. "The Sequence of Development of Some Early Mathematics Behaviors." Child Development 42: 1767-1778.

Wertheimer, M. 1938. "Number and numerical concepts in primitive peoples." In W. D. Ellis (ed.), A Sourcebook of Gestalt Psychology. New York: Harcourt, Brace & World.

Whitehead, A. N. and B. Russell. 1910-13. Principia mathematica. Vols. 1-3. Cambridge, England: Cambridge University Press.

ABOUT THE AUTHOR

CHARLES J. BRAINERD, Professor of Psychology, The University of Western Ontario, obtained his Ph. D. in developmental and experimental psychology in 1970. Since that time, he has contributed numerous articles to both educational and psychological journals. His work on number concepts appeared in Scientific American (March 1973).

Dr. Brainerd currently serves on the grant review committees of several organizations that fund research in the area of human number concepts (National Science Foundation, National Research Council, Canada Council, National Institute of Mental Health). Besides serving as editor of Journal of Child Development, he also serves as consultant to several publishers in the area of mathematics curriculum development. He was recently elected president of Society for Research in Child Development.